WASTEWATER ORGANISMS

A COLOR ATLAS

WASTEWATER ORGANISMS

A COLOR ATLAS

SHARON G. BERK
JOHN H. GUNDERSON

LEWIS PUBLISHERS
Boca Raton Ann Arbor London Tokyo

Library of Congress Cataloging-in-Publication Data

Berk, Sharon G.
Wastewater organisms : a color atlas / Sharon G. Berk and John H. Gunderson.
p. cm.
Includes bibliographical references and index.
ISBN 0-87371-623-X
1. Sewage—Microbiology—Atlases. I. Gunderson, John H. II. Title.
QR48.B47 1993
576—dc20 93-22831
CIP

International Standard Book Number 0-87371-623-X
Library of Congress Card Number 93-22831
Printed in the United States of America 1 2 3 4 5 6 7 8 9 0
Printed on acid-free paper

Preface

The organisms of the treatment system are key players in an effective wastewater treatment process. Upon close inspection of the wastewater, a variety of organisms can be observed having different sizes, shapes, life cycles, and roles in the processes. This book is intended to serve as a guide to the organisms commonly found in wastewater treatment facilities. It should be of interest to plant operators and other personnel, as well as to those interested in an overview of the organisms. It should aid the novice in identifying organisms, usually to the genus level. For those interested in species-level identification, more detailed techniques are given in the literature cited.

Most of the samples examined in the preparation of this book were taken from treatment plants in Tennessee, except for the parasites, which were collected from all across the U.S. The parasite photographs were kindly contributed by Dr. M. D. Little of Tulane University. Although most of the samples were from Tennessee, organisms found there have been reported by others from a variety of sites in the U.S. and Europe. The authors have not, however, found all species reported by other investigators. Fungi, for example, were not observed often enough in the samples examined to photograph; yet they are occasionally found in certain types of treatment processes.

Living specimens rather than fixed or stained specimens were photographed using differential interference contrast or phase contrast microscopy, often combined with a flash system. Although differential interference contrast is not usually acquired for non-research labs, it can show more details of living cells than other optical methods. It is our intention to provide the reader with the best view of such details of living organisms. This information may be useful in interpreting the structures of organisms viewed under the more commonly used brightfield or phase contrast techniques. The three-dimensional appearance of objects viewed with differential interference contrast optics also helps the viewer obtain a better understanding of the form and dimensions of the organisms. Although we chose to photograph living organisms, the chapter on enumeration provides literature and techniques for fixing and staining, which are often required for identification to the species level.

Sharon G. Berk
John H. Gunderson

Acknowledgments

We are grateful to the following individuals who have kindly volunteered to help collect and send samples from their treatment plants:

Dr. Isaac Barda, Metro Nashville Water Service, Central Lab, Nashville, TN
Mr. David Canizaro, McCullough Environmental Services, Athens, TN
Mr. Donald Dolezal, Lincoln, NE
Ms. Donna Howell, Water Quality Control Center, Murfreesboro, TN
Mr. Mike Jones, Municipal Corporation, Jefferson City, TN
Mr. Mark Kirk, Cookeville Wastewater Treatment Plant, Cookeville, TN
Mr. Ken Pate, Municipal Corporation, Jefferson City, TN
Mr. Larry Poteet, Baxter Wastewater Treatment Plant, Baxter, TN
Mrs. Peggy Smith and Mr. Marvin Smith, of Alexandria, TN
Mr. Jim South, Cookeville Wastewater Treatment Plant, Cookeville, TN
Ms. Linda Sprouls, Monterey Sewer Plant, Monterey, TN

We thank Dr. V. Dean Adams, former Director of the Center for the Management, Utilization and Protection of Water Resources at Tennessee Technological University, for encouraging us to work on this book. We also thank Dr. Dennis George, Associate Director of the Center for the Management, Utilization, and Protection of Water Resources, for helping us with questions concerning treatment processes. We are especially grateful to Dr. Hollings Andrews of the Biology Department of Tennessee Technological University for identifying the algae for us. Finally, we are very indebted to Dr. M. Dale Little of the School of Public Health and Tropical Medicine of Tulane University for contributing nearly all the photographs of parasites from sludge.

Dedication

We dedicate this book to the memory of Mr. Mark Kirk, who was a plant operator at the Cookeville Wastewater Treatment Plant. Mark helped us get started on the project, and many of the photographs in the book are from samples that he helped us collect. He was a fine gentleman and colleague who took his work seriously, yet had a cheerful attitude. He enjoyed conversing with us about the organisms, and he was also an amateur photographer. We are grateful for his help, and we will miss him.

About the Authors

John Gunderson received a B.S. in zoology and Russian from the University of Nebraska and his Ph.D. in zoology from the University of California at Berkeley. He held a postdoctoral position at the National Jewish Center for Immunology and Respiratory Medicine in Denver, Colorado in the laboratory of Mitchell Sogin where he used ribosomal RNA sequences to study protistan phylogeny. He also worked for a year in the Center for Molecular Evolution at the Marine Biological Laboratory in Woods Hole, Massachusetts before becoming an adjunct assistant professor in the Center for Management, Utilization and Protection of Water Resources at Tennessee Technological University. His research interests are in the field of protistan evolution. He serves on the Board of Reviewers for the *Journal of Eukaryotic Microbiology* and is a member of the Awards Committee of the Society of Protozoologists.

Sharon Berk received a B.S. in biology from the University of Illinois, a M.S. in zoology from the University of Maryland, and a Ph.D. in microbiology from the University of Maryland. She spent five years in the Department of Environmental Sciences of the University of Virginia as a research assistant professor. In 1987 she joined the faculty of the Center for the Management, Utilization and Protection of Water Resources at Tennessee Technological University, where she is currently a professor of biology. Her research interests lie in various aspects of environmental microbiology and include studies of protozoan-bacterial trophic interactions, and the development of toxicity tests using protozoa. Her recent projects include interactions of protozoa and *Legionella*, the bacterium that causes Legionnaires' Disease, and the use of artificial wetlands for wastewater treatment. She serves on *ad hoc* review panels of the National Institutes of Health and on the Standard Methods Committee for development of *Standard Methods for the Examination of Water and Wastewater*. She has been a long-time member of the Society of Protozoologists and the American Society for Microbiology. Courses she has taught include pollution microbiology, marine microbiology, microbial ecology and microscopy.

Table of Contents

Chapter I. Treatment Processes and Organisms 1

- Treatment Processes 3
- Bacteria 3
- Algae 4
- Protozoa 4
- Parasites 4
- Invertebrates 5
- References 5

Chapter II. Observation and Enumeration of Organisms 7

- Bacteria 9
  - Apparatus 9
  - Reagents 9
  - Procedure 10
- Algae 10
- Protozoa 10
- Invertebrates 11
- Parasites 12
  - Procedure for the Examination of Wastewater Sludge or Sludge Product for Helminth Eggs 13
  - Procedure for the Examination of Wastewater Sludge for Protozoan Cysts or Oocysts 16
- References 17

Chapter III. Photographs and Captions for Organisms

Glossary 19

Index of Genera 23

CHAPTER 1

Treatment Processes and Organisms

TREATMENT PROCESSES

Aerobic biological processes utilize microorganisms, primarily bacteria, to degrade colloidal or fine particulate matter and dissolved organics, resulting in carbon dioxide, water, and new cell biomass. Such processes produce flocculent settlable solids that can be removed in sedimentation tanks. Two of the most commonly used biological treatments include the activated sludge process and the trickling filter process.[1] The activated sludge process is most often used for large treatment systems. Most samples for the present study were collected from wastewater treatment plants utilizing activated sludge processes, and a few were from those using trickling filter systems.

In general, as sewage enters a wastewater treatment plant, it will initially be screened to remove coarse solids, then transferred to a grit chamber to remove sand and abrasive coarse material. From there, it is transferred to a primary settling tank to further remove other coarse solids. The portion of the contents of the primary settling tank containing fine particulates and dissolved organics is transferred to aeration tanks, which may be aerated by a variety of methods depending on the treatment plant. In the aeration tanks, much of the biological degradation takes place, and diverse communities of microorganisms thrive. The effluent from aeration tanks flows into another settling tank (or clarifier) to allow settling of flocculent settlable solids (activated sludge) produced during aeration. Some of the settled biological residuals from the clarifier are reintroduced into the aeration tank to ensure that an active microbial population is present. The process has become known as the activated sludge process because of the production of masses of active microorganisms produced and reintroduced during the treatment process.

Trickling filters consist of circular beds containing rocks or sheets of corrugated plastic over which the wastewater flows. Microorganisms attach to the surfaces of the rocks or plastic and degrade organic matter in the wastewater.

Residual solids (sludge) removed from primary sedimentation tanks can be digested anaerobically. It is placed in sealed tanks that may or may not be mechanically mixed. In a standard-rate digester, the contents are not usually heated or mixed, but in a high-rate process, the contents are heated and mixed. Heating will help destroy pathogens. Biological residuals from clarifiers are often digested aerobically, although it may be more costly than anaerobic digestion. For further details on these and other treatment processes, see Metcalf and Eddy, Inc.[1]

BACTERIA

The majority of bacteria present in activated sludge are heterotrophic, Gram-negative, motile rods.[2] Many are brought in with feces and may include human pathogenic bacteria. Their role in activated sludge is to oxidize organic waste to carbon dioxide and water and to produce flocculent settlable solids. Some may be chemolithotrophic and oxidize ammonia *(Nitrosomonas)*, nitrite *(Nitrobacter)*, and hydrogen sulfide *(Thiothrix)*. Filamentous bacteria such as *Sphaerotilus* may be found to cause bulked sludges, i.e., biological solids with poor settling characteristics and compactability.[1] Such filamentous bacteria have been referred to as "sewage fungi".

Anaerobic bacteria are active in anaerobic sludge digesters. They may originate from fecal material, land drainage, stormwater, and biological industrial effluents.[3] The role of these bacteria is to degrade sludges, which usually contain proteins, fats, and polysaccharides, into end products such as methane (which can be captured and used as an energy source in the treatment plant), carbon dioxide, water, and new bacterial biomass. Anaerobes include obligate anaerobes (those unable to live in the presence of free oxygen), facultative anaerobes (those able to switch from aerobic to anaerobic metabolism), and micro-aerophilic forms (those growing best in low oxygen concentrations). Most of the sulfide produced from sewage is from bacteria that reduce sulfates, such as *Desulfovibrio desulfuricans*. Another example of an anaerobic genus is *Methanobacterium*, which is responsible for methane production.

Identification of bacteria, even to the genus level, is tedious, requiring many biochemical (metabolic) tests or genetic analysis. The field of bacterial taxonomy is in flux, particularly with the appearance and use of ribosomal RNA analyses. It is virtually impossible to identify bacteria simply by observation under the microscope, and an educated guess is all that can be made based on appearance of cells. Very few have unusual characteristic morphologies that set them apart from the others. *Zoogloea,* for example, forms clusters of finger-like projections that few others do. Only a small proportion of all species are filamentous or contain sulfur granules. Most are small rods or curved rods. The photographs in this book serve primarily to demonstrate the variety of morphological types and sizes of bacteria encountered in wastewater treatment systems. Included in this section are the cyanobacteria, often still called blue-green algae. On the basis of cell structure, presence of murein, 70S ribosomes, and other features, these organisms are prokaryotes.[4] Cyanobacteria such as *Oscillatoria* photosynthesize as green plants do, evolving oxygen. The cyanobacteria of wastewater treatment systems can often be found on clarifier walls and trickling filters where they are exposed to sunlight.

ALGAE

Algae are chlorophyll-bearing plants that are distinguished from other chlorophyllous plants by their modes of sexual reproduction.[5] Some algae are unicellular, while others may be multicellular. Some are suspended in the water column and are termed planktonic, while others may be attached to substrates such as rocks or as in the case of treatment facilities, clarifier walls or trickling filters.

Many species of algae are photoautotrophic, i.e., they get energy from light and all other nutrients from inorganic compounds such as carbon dioxide and ammonia. Some also require other growth factors such as vitamins and are termed photoauxotrophic. A few species are facultatively heterotrophic, which means they can switch to a mode of nutrition that utilizes organic compounds. Still a few others, in addition to photosynthesizing, may ingest solid particles.

In wastewater treatment systems, algae grow on the nutrients in the water and get energy from sunlight. Generally they are found in highest numbers on surfaces such as trickling filters or walls of clarifiers, where the biologically treated waters trickle over the top of the clarifiers. In such an environment are found many filamentous or branched forms attached to the substrate. Algae play a minor or insignificant role in wastewater treatment. They may even cause problems of clogging or eutrophic blooms after discharge of treated wastewater to receiving waters. However, they are some of the loveliest specimens to observe with a microscope.

PROTOZOA

Several groups of protozoa are represented in wastewater treatment plants. These include flagellates, ciliates, amoebae, and sporozoa. The sporozoa are parasites of humans and animals and therefore are included in a separate section on parasites. The taxonomic division of predominately unicellular eukaryotes into algae and protozoa, and the protozoa into flagellates, ciliates, amoebae, and sporozoa does not reflect their true evolutionary relationships. However, this traditional scheme is followed here, solely for convenience. It is the scheme that will be found in the references cited. In several texts on wastewater treatment, suctorians have been regarded as a separate class from ciliates. However, they are ciliates that lack cilia in the adult stage, but possess cilia in the larval stage.

Protozoa are abundant and diverse in activated sludge, where they feed primarily on bacteria and organic nutrients. Protozoa demonstrate a variety of trophic (feeding) modes. Some feed on bacteria, some on algae, some on other protozoa, and some feed on a mixture of food types such as algae and bacteria. They may originate from various components of sewage or from material of the conduits collecting sewage and runoff. Many species form cysts that can be airborne or picked up by birds and aquatic insects. Several anaerobic species have also been observed in Imhoff tanks and digesters and are thought to feed on sulfur bacteria.

The primary role of protozoa is to clarify the effluent through predation on bacteria. In addition, protozoa may cause flocculation, which is thought to result from secretion of a mucous-like substance from certain species.[6] However, protozoan-induced flocculation may not be significant in clarification of the effluent. Protozoa are capable of utilizing dissolved organic substrates; however, the extent to which this affects the effluent is unknown.[6] In addition to removal of bacteria, predation or grazing of bacteria by protozoa plays another role. Grazing on bacteria, especially on surfaces, has been shown to enhance bacterial community metabolism. By cropping bacteria, the protozoa remove senescent bacteria and maintain the population in a youthful physiological state increasing the rate of bacterial assimilation of organic matter.[7]

The structure of the protozoan community is believed to reflect the quality of the effluent. Ciliates, in particular, have been considered indicators of a good treatment process.[6] However, it should be remembered that incoming wastewater may derive from many different sources and therefore contain different nutrients or toxic compounds. Protozoa (and other organisms) may adapt to a particular chemical composition of wastewater, resulting in different community structures, depending on the composition of wastewater. In samples from various treatment plants examined for this book, the dominant species differed from plant to plant; yet no plant reported a poor quality effluent. More research on community structure and quality of treatment processes is necessary.

PARASITES

Parasites of humans and animals are different from many other sewage organisms, in the sense that they are not part of an active community of organisms interacting with the sewage and other organisms, but rather, stages that just exist in the sewage as resistant forms, usually as eggs or cysts. For these reasons, the parasitic protozoa and other parasitic invertebrates have been placed in a section of this book separate from the free-living protozoa and invertebrates.

Parasites enter the treatment system with feces of humans and animals. Although most are killed during sludge treatment, some such as ascarids can survive the harsh environment of the treatment process. Anaerobic digestion does, however, destroy more parasites than aerobic digestion of sludges. Some parasitic forms can even resist the concentrations of chlorine used as a disinfectant prior to release of a treated effluent.[8] In general, exposure of raw sewage to treatment processes shortens the survival time of the parasites. The parasites do not play a role in the treatment processes. More details concerning characteristics and life cycles of sewage parasites are provided in the photograph section of this book.

INVERTEBRATES

Two of the most common groups of invertebrates encountered by the authors were the rotifers and nematodes. Rotifers are multicellular invertebrates usually between 100 and 500 μm. Most are found in freshwater. The term "rotifer" (= "wheel-bearer") derives from the fact that most of the species have a ciliated anterior end (corona) that resembles a pair of rotating wheels due to the synchronous beating of the cilia. Natural populations are almost all females, and reproduction is parthenogenetic.[9] Males are unknown for many species, and in those species in which they occur, males are rare and short-lived. Nematodes are roundworms, usually less than 1 cm long. They are difficult to indentify and, as a group, are poorly known.[9] They move with a whip-like motion. Rotifers are most common in activated sludge, and nematodes are usually found in trickling filter material or in treatment plants with percolating filters. The roles of rotifers and nematodes are not clear. Both serve to consume and digest particulate matter and other smaller organisms. Annelids, with similar roles, are found among filamentous material.

Tardigrades (water bears) were encountered in only one treatment plant, and gastrotrichs were found less frequently than rotifers and nematodes. Fox et al.[8] found copepods and cladocerans in sewage of the Chicago area; however, these groups are not considered to be normal inhabitants of a wastewater system. These groups were not observed in any samples of the present study. More details on structure and habits are provided in the photograph section.

REFERENCES

1. Metcalf and Eddy, Inc. *Wastewater Engineering: Treatment, Disposal, Reuse* (New York: McGraw-Hill, 1979), pp. 468–572.

2. Pike, E. B. "Aerobic Bacteria," in *Ecological Aspects of Used-Water Treatment*, C. R. Curds and H. A. Hawkes, Eds. (London: Academic Press, 1975), p. 16.
3. Crowther, R. F. and N. Harkness. "Anaerobic Bacteria," in *Ecological Aspects of Used-Water Treatment*, C. R. Curds and H. A. Hawkes, Eds. (London: Academic Press, 1975), pp. 65–92.
4. Stolp, H. *Microbial Ecology: Organisms, Habitats, Activities* (Cambridge: Cambridge University Press, 1988), pp. 61–68.
5. Bold, H. C. and M. J. Wynne. *Introduction to the Algae* (Englewood Cliffs, NJ: Prentice-Hall, 1985), p. 1.
6. Curds, C. R. "Protozoa," in *Ecological Aspects of Used-Water Treatment*, C. R. Curds and H. A. Hawkes, Eds. (London: Academic Press, 1975), pp. 203–268.
7. Johannes, R. E. "Influence of Marine Protozoa on Nutrient Regeneration," *Limnol. Oceanogr.* 10:434–442 (1965).
8. Fox, J. C., P. R. Fitzgerald, and C. Lue-Hing. *Sewage Organisms: A Color Atlas* (Chelsea, MI: Lewis Publishers, 1981), pp. 13–23.
9. Pennak, R. W. *Fresh-Water Invertebrates of the United States* (New York: John Wiley & Sons, 1989).

CHAPTER 2

Observation and Enumeration of Organisms

BACTERIA

Living bacteria can be visualized quite well by phase contrast microscopy, using a 40× or 100× objective. Some are motile and may swim across the field of view, while others are not motile but may appear to vibrate or drift under the coverslip. They are almost invisible under brightfield optics if they are not stained. Various procedures are available for enumerating viable (living) bacteria. Several of these procedures include growing a diluted sample on solid agar media. This may result in an underestimation of the total number of living bacteria, because the media may not be suitable for the requirements or tolerances of all species present in the water sample. Also, some bacteria may be slow growers. Plate count techniques for specific bacteria and total heterotrophic bacteria are outlined in *Standard Methods for the Examination of Water and Wastewater.*[1]

Other means of assessing viability are slightly more complicated and involve incubation of water samples containing bacteria with chemicals such as a tetrazolium chloride compound that becomes reduced to a formazan residue within respiring cells.[2] The reddish formazan compound can be seen within cells using brightfield microscopy. Other techniques include incubation with radioactive substrates followed by autoradiography[3] or with nalidixic acid, which inhibits cell division resulting in elongated cells.[4]

One of the simplest and most widely used techniques for enumerating total bacteria is epifluorescence microscopy after staining with a fluorochrome such as acridine orange.[5] These techniques, however, do not distinguish between living and dead cells. Training by an experienced technician may be necessary to learn to distinguish between bacteria and other fluorescing debris and artifacts. The following acridine orange method is modified from *Standard Methods for the Examination of Water and Wastewater,*[1] and it was originally described by Hobbie et al.[5]

Apparatus

- Microscope with an appropriate illuminator for epifluorescence (a Hg lamp or 100 W halogen lamp for certain microscopes and filters) and with a flat field 100× oil immersion objective to give a total magnification of 1000×
- Counting graticule, ocular lens micrometer calibrated with a stage micrometer
- Filters, including excitation filters (wavelength 455 or 490 nm), dichroic mirror or beam splitter (wavelength 510 nm), and barrier filter (520 nm, long pass)
- Blender or vortex mixer
- Filtration unit suitable for use with 25 mm diameter membrane filters
- Membrane filters, polycarbonate, 25 mm diameter, 0.2 µm pore size (purchase black nonfluorescent or prepare by soaking in Irgalan black [2 g/L in 2% acetic acid] for 24 hr, then rinse in water and air-dry)
- Syringes or sterile pipets
- Test tubes, glass, to hold 10 mL

Reagents

- Phosphate buffer: dissolve 13.6 g KH_2PO_4 in water and dilute to 1 L. Adjust to pH 7.2 if necessary; filter through 0.2 µm membrane filter.
- Fixative, 5.0% (w/v) glutaraldehyde in phosphate buffer. Prepare fresh daily.
- Fluorochrome, 0.1% acridine orange in phosphate buffer, filtered through 0.2 µm pore sterile filter.
- Immersion oil, low fluorescing, Type B.

Procedure

1. Add 9 mL sample to test tube containing 1.0 mL fixative. Fixed samples can be stored for several weeks at 4°C.

2. Mix sample using blender or vortex mixer, then make 10-fold dilutions in phosphate buffer as necessary.

3. Place 1 mL sample or dilution in the filter tower with the black filter in place (shiny side of filter up).

4. Add 1 mL acridine orange and allow 2 min for staining, then add about 3 mL filtered buffer to promote even distribution of cells. Alternatively, add sample and acridine orange in test tube and stain for 2 min before pouring into the filter tower.

5. Filter with a vacuum (about 13 kPa).

6. Wash with 2 mL of buffer and filter again.

7. Remove filter with forceps and allow to air dry for about a minute.

8. Place the filter on a small drop of immersion oil applied to a glass microscope slide.

9. Add another small drop of immersion oil to the filter surface without touching the applicator to the filter.

10. Gently place a glass coverslip on top of the filter and tap the coverslip to distribute oil evenly over the filter and to reduce the thickness of the preparation. Stained samples can be stored in the dark for several months without significant loss of fluorescence.

11. Examine at least 10 randomly selected fields using the 100× oil immersion objective to determine whether the sample needs to be diluted and that the cells are evenly distributed.

12. Count the number of cells in at least 20 large squares of the calibrated counting graticule.

13. Calculate the number of graticule squares on the effective area of the filter under the 100× objective.

14. Determine the average number of cells per square, and this will give the average number of cells per filter. The total cells per milliliter = (average no. cells per square) × (squares per filter) × (dilution factor)/(sample volume passed through the filter, in milliliters).

ALGAE

Algae can be visualized by brightfield microscopy better than bacteria can, although phase contrast is very good, especially for observing flagella of flagellated algae. Since many are not motile, they can be observed for a long period under the coverslip. To preserve samples for enumeration, Lugol's iodine solution is most suitable. The following procedure is taken from *Standard Methods for the Examination of Water and Wastewater*:[1]

Lugol's solution: Dissolve 20 g of potassium iodide (KI) and 10 g iodine crystals in 200 mL glacial acetic acid. To preserve a sample, add 0.3 mL Lugol's solution to 100 mL sample and store in the dark. For long term storage, add 0.7 mL Lugol's solution to 100 mL sample; then after 1 hr, add buffered formaldehyde to a minimum of 2.5% final concentration.

Since algae are so varied in size compared with bacteria, they may be enumerated under different conditions, depending on size. For example, a Sedgwick-Rafter counting chamber, which holds 1 mL samples, can be used for the larger cells visible with low magnification (up to 200×). The chamber is too thick for higher magnification objectives. For intermediate size algae, a Palmer-Maloney counting chamber can be used. It holds 0.1 mL samples and can be used with higher magnification objectives such as 40×. For small algae, oil immersion objectives might be used. There are several methods for observing and counting cells using high magnification. For details on counting techniques and calculations, see *Standard Methods for the Examination of Water and Wastewater*,[1] section 10200 F. In general, if the entire contents of the chamber cannot be scanned, random fields or fractions of the chamber volume can be examined for number of cells, and calculations made accordingly.

PROTOZOA

The best way to observe living protozoa (and other small motile organisms) is with differential interference contrast microscopy or phase contrast microscopy. The latter is more common and less expensive. Differential interference contrast (often referred to as Nomarski interference) is optically more complex than phase contrast methods. For a detailed explanation of principles of differential interference microscopy, consult texts on microscopy such as *Microscopy and Photomicroscopy, A Working Manual*.[6] These optical systems allow visualization of many structures including vacuoles, nuclei, cilia, or flagella and details of surface patterns on certain protozoa. The pattern of movement shown by organisms is also characteristic of certain groups. For photographing protozoa, a flash unit is most useful in stopping movement including ciliary or flagellar motion. If a flash unit is not available, there are alternatives; however, they are not as effective. Slowing down of protozoa may be accomplished by adding chemicals to the specimen on the slide. A methylcellulose compound, Protoslo (Carolina Biological Supply Co., Burlington, NC) increases the viscosity of the medium, slowing down the organisms. It can be messy, however. Other agents are used to narcotize or immobilize cells. These include nickel sulfate (0.1% w/v) and nickel or copper chloride (5 to 10 mMol/L). Maintaining a minimal volume of water under a coverslip may help trap larger protozoa. Fixatives such as Lugol's solution or formalin will kill the protozoa, and organisms respond differently to different fixatives. Most cells will become distorted upon fixation. Many different fixatives useful for protozoa are described in *An Illustrated Guide to the Protozoa*[7] and *Protocols in Protozoology*.[8] Volatile fixatives should be used under a hood. Special staining techniques are available to visualize ciliary patterns and organelles. One of the most useful stains is a silver impregnation technique using Protargol. This is an excellent stain, but not simple. The procedure for the Protargol silver stain is outlined in detail in *An Illustrated Guide* and in *Protocols in Protozoology*.[8] Other simpler stains are also included in *An Illustrated Guide*,[7] as well as in *Sewage Organisms: A Color Atlas*.[9] The following stain, taken from *An Illustrated Guide to the Protozoa*[7] (developed by Borror[10]), is recommended for use in examining many organisms quickly, especially ciliates:

Borror's Nigrosin-HCl-Formalin Stain

1. Transfer cells by pipet to fresh Champy's fixative for 5 min.
2. Transfer to distilled water to wash.
3. By pipet, place a drop about 1 cm diameter on a slide.
4. Add, by pipet, from 2 to 3 cm above the slide, 2 to 3 drops of nigrosin–HCl–formalin fixative (12 parts Nissenbaum's solution; 1 part Deroux-–Faidy's solution— see below).
5. Remove excessive fixative by pipetting.
6. Repeat steps 4 and 5.
7. Dehydrate in isopropanol series, 70, 95, 100%, 5 min each.
8. Clear in xylene, 5 min.
9. Mount in mounting medium under no. 1 coverslip.

Champy's Fixative	
7 parts	Chromic acid (1%)
7 parts	Potassium dichromate (3%)
4 parts	Osmium tetroxide (2%) (use inside hood)
Nissenbaum's Solution	
10 mL	$HgCl_2$ (saturated aqueous soln.)
2 mL	Glacial acetic acid
2 mL	Formalin (40%)
10 mL	*t*-Butanol
Deroux–Faidy's Solution	
20 mL	Formalin (40%)
4 g	Nigrosin (aqueous soln.)
100 mL	Distilled water

Amoebae without tests are very difficult to identify accurately.[11] Some useful nuclear characteristics can only be seen after the amoebae are fixed and stained. The size of the organism and the shape of the pseudopods are important characteristics, but change with environmental conditions. The pressure of a coverslip or heat and light of the microscope can cause the amoebae to retract pseudopods and round up. As a result, identification of amoebae usually necessitates raising cultures of them under controlled conditions. Even identifying the family to which a particular amoeba from an environmental sample belongs can be a challenge.

Enumeration of protozoa must be done with fixed samples. The counting procedures used for algae are appropriate for protozoa, although Lugol's iodine solution cannot be used if organisms are to be identified to species. Other fixatives such as Bouin's[7] are better in such cases, because organisms can then be stained in the manner necessary for identification. Random microscopic fields or the entire volume of some counting chambers such as a Palmer-Maloney counting cell can be examined for numbers of protozoa. Samples may need to be concentrated. Gentle centrifugation (approx. $125 \times g$) in 10 mL Kolmar tubes (Fisher Scientific, Norcross, GA) can be used for samples up to 10 mL. Settling chambers combined with an inverted microscope may be used for larger volumes of fixed samples. A gentle technique for concentrating plankton has been developed by Dodson and Thomas[12] and can be modified for protozoa. Basically this consists of a beaker of unconcentrated sample into which is lowered a tube, covered at the bottom with netting of appropriate mesh to filter out the protozoa of interest. As the tube fills with water (filtered), the water inside the tube can be removed with a pipet and discarded. For an illustration, see *Standard Methods for the Examination of Water and Wastewater,*[1] method 10200 G. Concentrated samples can then be fixed and enumerated using various counting chambers.

INVERTEBRATES

Prior to fixation, concentration of small invertebrates can be accomplished by the gentle reverse filtration technique of Dodson and Thomas,[12] as described above for zooplankton. Counting chambers such as the Sedgwick-Rafter or Palmer-Maloney counting cells can be used for enumerating the small invertebrates.

Pennak[13] described various methods for preparing and preserving invertebrates. For loricate rotifers, he recommends fixing them directly in 10% formalin, which will kill them and cause the soft parts of the animal to contract strongly, leaving the lorica well exposed for examination. Identification of loricate species depends largely on details of the lorica.

For those without loricas, it is desirable to have the organism relaxed and uncontracted. These species must be narcotized before killing and fixing. Pennak[13] recommends chloretone, 2% benzamine lactate, 2% butyn, or 2% hydroxylamine hydrochloride. Some species without loricas, especially bdelloid rotifers, are too delicate for narcotization, and for such organisms, the specimens can be flooded with boiling water in a watch glass or depression slide. This treatment will leave most organisms in a well-expanded condition. The organisms should then be fixed with formalin or other fixative. After killing and fixing, all specimens can be stored in 2 to 5% formalin with 2% glycerin. Eosin can be added to aid in visualizing and studying them.

For gastrotrichs, an excellent fixative is osmic acid. Specimens in a drop of water on a slide can be inverted over the mouth of a bottle of osmic acid. Other good fixatives include saturated mercuric chloride, 10% formalin, and Bouin's fixative.[13]

Nematodes are so active that they must be narcotized or killed before they are examined for identification under the microscope. Narcotization can be accomplished with chloroform, chloral hydrate, magnesium sulfate, or other narcotic.[13] Hot saturated mercuric chloride and hot 70 to 80% alcohol are good fixatives, although cold 85% alcohol and 5% formalin are reliable for the field. Also, killing in a 57°C water bath for 10 min can give good results.[13]

PARASITES

The following detailed procedures for detection of parasites in sludge were developed and provided by Dr. Little of the School of Public Health and Tropical Medicine, Tulane University Medical Center.

Procedure for the Examination of Wastewater Sludge or Sludge Product for Helminth Eggs

Reagents

- 1% solution of "7X" (Limbro) (v/v), an anionic detergent
- $MgSO_4$ (USP) solution, specific gravity 1.20, adjusted with hydrometer
- $MgSO_4$ (USP) solution, specific gravity 1.06, adjusted with hydrometer
- $NaNO_3$ solution, saturated, specific gravity 1.375
- Culture fluid, 0.5% formalin (5 mL formalin and 995 mL H_2O)
- Aqueous iodine solution (1 g iodine crystals, 2 g potassium iodide, 100 mL dist. H_2O)
- Prosil 28[R], Organosilane for coating glassware. All glassware should be treated according to instructions and at intervals recommended by manufacturer

Materials

- Beakers, 1000 mL, tall form, Berselius, Pyrex
- Beakers, 100 mL, low form, Pyrex
- Funnel, glass, short stem, 50 mm diameter
- Funnel, powder filling, 150 mm diameter, Nalgene
- Centrifuge tubes, conical, polypropylene, 50 mL
- Centrifuge tubes, conical, Pyrex, 15 mL
- Pipettes, Pasteur, disposable, 5 $^3/_4$ in.
- Coverglasses, 22 × 30, 24 × 40, 24 × 50
- Microscope slides, glass, 3 × 2 in.
- Sieve, 20 mesh, 7 in. diameter
- Sieve, 400 mesh stainless-steel screen attached to bottom of cut-off stainless-steel funnel (5 in. diameter)
- Hydrometer, 1.0 to 1.3 sp. gr.
- Culture dishes (60 × 15 mm glass Petri dishes)
- Wash bottles, plastic, 500 mL
- Spray bottles, 16 oz
- Parafilm[R], 4 in. wide roll

Equipment

- Centrifuge, with head and cups for 50 mL tubes
- Centrifuge, clinical model, with head for 15 mL tubes
- Blender, Waring with 40 oz glass container
- Magnetic stirrer
- Vortex mixer
- Ultrasonic cleaner, 11.5 ×5.5 × 6 in. (for cleaning sieves)
- Microscope, compound, binocular

Procedure

1. For liquid sludges, take a measured volume (100 mL or more to contain at least 5 g dry solids), place in blender container, add about 200 mL of water, and blend for 1 min at high speed. For dewatered sludges, take a weighed sample (about 30 g), place in 200 mL of water in a beaker, and let soak overnight. Then transfer sample to blender and blend for 1 min at high speed.

2. Pour blended sample into a 1000 mL tall form beaker, and using wash bottle, thoroughly rinse blender container into beaker. Add 1% "7X" to 900 mL.

cont.

3. Let settle for 4 hr (or overnight). While settling, the solution may need to be stirred occasionally to ensure that any material floating on the surface will settle.

4. After settling, pour off the supernatant, transfer to blender, add water to 300 mL, and blend again for 1 min at high speed.

5. Transfer to tall beaker, add "7X" to 900 mL, and allow to settle for 2 hr. Pour off supernatant and discard.

6. Add 300 mL of "7X" and stir for 5 min on a magnetic stirrer.

7. Strain homogenized sample through a 20 mesh sieve into a beaker. The sieve is placed on a funnel over a tall beaker, and the sample is washed through the sieve with the aid of "7X" spray from a spray bottle.

8. Add "7X" to 900 mL and allow to settle for 1 to 2 hr.

9. Decant and discard supernatant; mix sediment and distribute equally to eight 50 mL centrifuge tubes. Thoroughly wash any sediment in beaker into tubes with the aid of a wash bottle. Bring volume in each tube to 50 mL with water.

10. Centrifuge for 5 min at 800 g and pour off supernatant from each tube and discard. (The packed sediment in each tube should not exceed 5 mL; if so, add water and distribute evenly among additional tubes; repeat centrifugation and decant supernatant.)

11. Add 10 to 15 mL of $MgSO_4$ solution (sp. gr. 1.20) to each tube, cap and mix for 15 to 20 sec on a Vortex mixer.

12. Add additional $MgSO_4$ solution to each tube to bring to 50 mL; centrifuge for 5 min at 800 g.

13. After centrifuge has come to a complete stop (without use of brake), pour the top 25 to 35 mL of supernatant from each tube through a 400 mesh sieve (sieve can be placed over a tall beaker).

14. Using the spray bottle, wash the excess flotation fluid and fine particles through the sieve.

15. Rinse the sediment collected on the sieve into a 100 mL beaker by directing a stream of water from a wash bottle onto the upper surface of the sieve.

16. After thoroughly washing the sediment from the sieve, transfer the suspension to the required number of 15 mL centrifuge tubes, taking care to rinse the beaker into the tubes.

17. Centrifuge the tubes for 3 min at 800 g, then discard supernatant.

18. If more than one tube has been used for the sample, transfer the sediment to a single tube, fill with water, and repeat the centrifugation.

19. Add 2 to 3 mL of culture fluid (0.5% formalin) to sediment, mix, and add culture fluid to fill tube.

20. Centrifuge for 3 min at 800 g, then pour off supernatant.

21. Add a few drops of culture fluid to sediment, mix with a clean pipette, and transfer to culture dish. Rinse tube and pipette with about 1 mL of culture fluid into culture dish. Add culture fluid, if needed, to have fluid completely cover bottom of dish (a depth of 3 to 4 mm). Place a sheet of Parafilm over Petri dish bottom to seal, then cover with Petri dish top.

22. Store culture in dark or subdued light at 22 to 28°C. Dishes should be agitated every few days to mix sediment and aerate fluid.

cont.

23. After a minimum of 28 days, the cultured sediment is transferred to one or more 15 mL centrifuge tubes, depending on the volume of sample, by placing a small glass funnel into the tube and rinsing the suspension into the tube.

24. Centrifuge at 800 g for 3 min, then pour off supernatant.

25. Add 10 mL of 10% commercial bleach to the sediment and mix thoroughly. Allow to stand for 5 min. (This decolorizes the outer shell of the eggs of *Ascaris* and *Trichuris* and permits the contents of the eggs to be observed more easily.)

26. Centrifuge at 800 g for 3 min, then pour off supernatant.

27. Add 10 to 15 mL of water, mix, and repeat step 26.

28. With pipette, mix sediment with the small amount of water remaining, transfer a portion of sediment to a microscope slide, and cover with an appropriate size of coverglass (fluid should not extend beyond edges of coverglass). Examine entire sediment.

29. Systematically examine each preparation under at least 100× magnification of the microscope. If necessary, check identity of objects under 400× magnification.

30. Count all of each type of parasitic helminth egg present (*Ascaris, Trichuris, Toxocara*, etc.) and record whether they contain a developed, motile larva (viable egg) or are undeveloped (dead or nonviable egg).

31. Report the numbers of viable eggs present on a dry weight basis, i.e., number of eggs per gram (or kg) of solids (TSS).

Modification of Procedure in Order to Detect Taenia *Eggs*

Taenia eggs have a specific gravity of about 1.25, i.e., higher than that of *Ascaris, Toxocara,* and *Trichuris* species and will not ordinarily be recovered by the procedure previously given, which uses a $MgSO_4$ solution with a specific gravity of 1.20. If it is suspected that *Taenia* eggs might be present in a sludge sample due to the nature of the population served by the wastewater treatment plant, the following additional procedure should be used. This procedure is a modification of the procedure of Owen as described in "Guidelines for Surveillance, Prevention and Control of Taeniasis/Cysticercosis."[14]

1. After step 13 (above), save the sediment in each of the 50 mL tubes after completely pouring off the remaining supernatant.

2. Add about 10 mL of a saturated solution of $NaNO_3$ (specific gravity of 1.375) to each tube, cap and mix for 15 to 20 sec on a Vortex mixer.

3. Add additional $NaNO_3$ solution to each tube to bring to 50 mL; centrifuge for 5 min at 800 g.

4. After centrifuge has come to a complete stop (without use of brake), pour the top 10 to 15 mL of supernatant from each tube into a 1000 mL beaker. (An alternative method is to aspirate the surface film from each tube and transfer to a separate centrifuge tube.)

5. Add water to the beaker to dilute the flotation solution to below the specific gravity of the eggs. Add water to nearly fill container. Allow mixture to settle for 1 hr.

6. Decant supernatant, transfer sediment to one or more 15 mL tubes as needed, centrifuge at 800 g for 2 min, and decant supernatant.

7. Transfer sediment to one or more microscope slides, as is required, cover with an appropriately sized coverglass, and examine directly under at least 100× magnification of the microscope. Check identity of egg-like objects under 400× magnification. The eggs of all *Taenia* species

are indistinguishable so that they should only be reported as "*Taenia* eggs".

8. Report the number of eggs present on a dry weight basis, i.e., number of eggs per gram (or Kg) of solids.

9. Since *Taenia* eggs are normally infective when they are passed into the sewage system, culturing to determine viability is not required. Determination of viability of eggs is not always possible, but viability is indicated when the hooklets of the hexacanth embryo within the egg shell are observed to be moving.

Procedure for the Examination of Wastewater Sludge for Protozoan Cysts or Oocysts

The resistant stages of the protozoan pathogens of humans and animals are occasionally observed in raw or untreated sludges, but not in treated sludges. During most sludge treatment processes, these stages either die or, since they are much lighter than helminth eggs, are separated from the sludge particles and end up in the effluent water.

Reagents, Materials, and Equipment

Same as above.

Procedure

1. For liquid sludges, take a measured volume (100 mL or more to contain at least 5 g dry solids), place in blender container, add about 200 mL of water, and blend for 1 min at high speed. For dewatered sludges, take a weighed sample (about 30 g wet weight), place in 200 mL of water in a beaker, and let soak overnight. Then transfer sample to blender and blend for 1 min at high speed.
2. Pour blended sample into a 1000 mL tall form beaker and using wash bottle, thoroughly rinse any sediment remaining in blender container into beaker. Add 1% "7X" to a volume of 900 mL.
3. Let settle for 4 hr (or overnight). While settling, the solution may need to be stirred occasionally to ensure that any material floating on the surface will settle.
4. After settling, pour off the supernatant, transfer to blender, add water to 300 mL, and blend again for 1 min at high speed.
5. Transfer to tall beaker, add "7X" to 900 mL, and allow to settle for 2 hr. Pour off supernatant and discard.
6. Add 300 mL of "7X" and stir for 5 min on a magnetic stirrer.
7. Strain homogenized sample through a 20 mesh sieve into a beaker. The sieve is placed on a funnel over a tall beaker, and the sample is washed through the sieve with the aid of spray of "7X" spray from a spray bottle.
8. Add "7X" to 900 mL and allow to settle for 3 or more hours.
9. Decant and discard supernatant; mix sediment and distribute equally to eight 50 mL centrifuge tubes. Thoroughly wash any sediment in beaker into tubes with the aid of a wash bottle. Bring volume in each tube to 50 mL with water.
10. Centrifuge for 5 min at 800 g and pour off supernatant from each tube and discard. (The packed sediment in each tube should not exceed 5 mL; if so, add water and distribute evenly among additional tubes; repeat centrifugation and decant supernatant.)
11. Add 10 to 15 mL of $MgSO_4$ solution (sp. gr. 1.060) to each tube, cap and mix for 15 to 20 sec on a Vortex mixer (use care not to cause splashing of mixture from tube).
12. Add additional amount of $MgSO_4$ solution to each tube to bring to 50 mL; centrifuge for 5 min at 800 g.
13. After centrifuge has come to a complete stop (without use of brake), pour the top 10

cont.

to 15 mL of supernatant from each tube into a 1000 mL beaker. (An alternative method is to remove the surface film and the top few milliliters of the flotation fluid from each tube with a pipette and transfer to a beaker.)

14. Add sufficient amount of water to beaker to dilute the flotation fluid to below the specific gravity of the protozoan cysts; add water to nearly fill beaker. Allow mixture to settle to 2 to 3 hr.

15. Decant supernatant, transfer sediment to one or more 15 mL conical centrifuge tubes, as needed, centrifuge at 800 g for 5 min, and pour off supernatant.

16. When ready to examine, add an equal volume of aqueous iodine solution to sediment in tube and transfer a drop of the sediment to a glass microscope slide. Cover with a coverglass and examine under the microscope. Slide should be scanned under 100× magnification, and when objects appearing to be protozoa are seen, they should be examined under 400× magnification or higher. Examine the entire sediment in the same manner.

17. Identification is made on the basis of size, shape, number, and type of nuclei; presence of chromatoidal bars, axonemes, etc., and comparison with the cyst and oocyst stages of different protozoan species as they are illustrated in atlases and books.

18. Quantification is done only in general terms, e.g., few, moderate, or many present.

Note: This procedure for the detection of the resistant stages of protozoa in sludges can be performed on the same sample that is to be examined for helminth eggs. In this case, when the packed sediment remaining after the flotation step using $MgSO_4$ solution with a specific gravity of 1.060 is completed (step 13), it can then be used in the procedure for detecting helminth eggs. Using this packed sediment, the examination for helminth eggs would start at step 11.

REFERENCES

1. *Standard Methods for the Examination of Water and Wastewater*, 18th ed. (Washington, D.C.: American Public Health Association, American Waterworks Association, Water Environment Federation, 1992).
2. Zimmerman, R., R. Iturriaga, and J. Becker-Birck. "Simultaneous Determination of the Total Number of Aquatic Bacteria and the Number Thereof Involved in Respiration," *Appl. Environ. Microbiol.* 36:926–935 (1978).
3. Meyer-Reil, L. -A. "Autoradiography and Epifluorescence Microscopy Combined for the Determination of Number and Spectrum of Actively Metabolizing Bacteria in Natural Waters," *Appl. Environ. Microbiol.* 36:506–512 (1978).
4. Kogure, K., U. Simidu, and Taga. "A Tentative Direct Microscopic Method for Counting Living Marine Bacteria," *Can. J. Microbiol.* 25:415–420 (1979).
5. Hobbie, J. E., R. J. Daley, and S. Jasper. "Use of Nucleopore Filters for Counting Bacteria by Fluorescence Microscopy," *Appl. Environ. Microbiol.* 33:1225–1228 (1977).
6. Smith, R.F. *Microscopy and Photomicroscopy, A Working Manual* (Boca Raton, FL: CRC Press, 1992).
7. Lee, J. J., S. H. Hutner, and E. C. Bovee. *An Illustrated Guide to the Protozoa* (Lawrence, KS: Society of Protozoologists, 1985).
8. Lee, J. J. and A. T. Soldo. *Protocols in Protozoology* (Lawrence, KS: Society of Protozoologists, 1992).
9. Fox, J. C., P. R. Fitzgerald, and C. Lue-Hing. *Sewage Organisms: A Color Atlas* (Chelsea, MI: Lewis Publishers, 1981), pp. 3–4.
10. Borror, A. C. "Nigrosin-HCl-Formalin: a Stain-Fixa-tive for Ciliates," *Stain Technol.* 43:273–274. (1968).
11. Page, F. C. *A New Key to Freshwater and Soil Amoebae* (Ambleside, Cambria: Freshwater Biological Association, 1988).
12. Dodson, A. N., and W. H. Thomas. "Concentrating Plankton in Gentle Fashion," *Limnol. Oceanogr.* 9:455 (1974).
13. Pennak, R. W. *Fresh-water Invertebrates of the United States* (New York: John Wiley & Sons, 1989).
14. World Health Organization. "Guidelines for Surveillance, Prevention and Control of Taeniasis/Cysticercosis" (Gemmel *et al.*, Eds., 1983).

CHAPTER 3

Photographs and Captions for Organisms

A list of references used for the information in this chapter is given at the end of the photograph section.

Bacteria

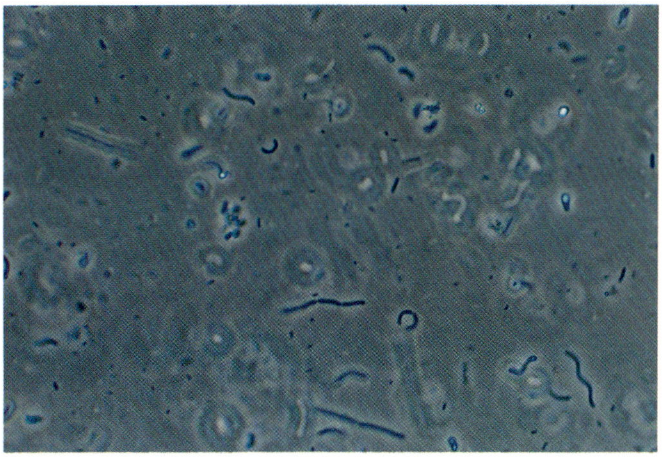

Figure 1. A variety of bacterial sizes and shapes can be seen in this photograph. They include rods, curved rods, chains of rods, spirals, and coccoid (spherical) shapes. Sample was aged for a few days from final clarifier, Cookeville, TN. Phase contrast.

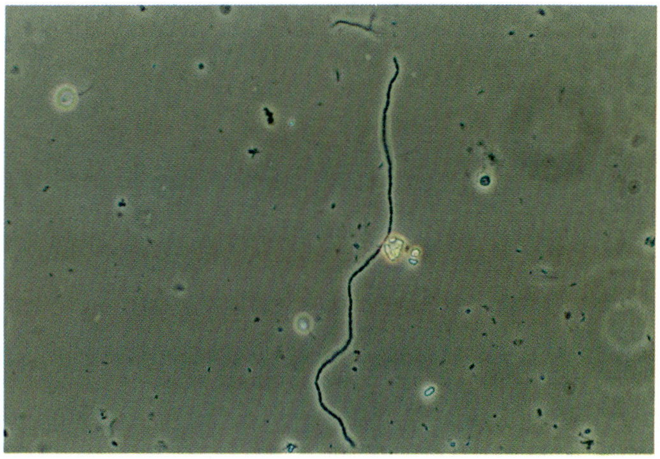

Figure 2. A long chain of rod-shaped bacteria. Chain is ≈150 μm long, from final clarifier, Cookeville, TN. Phase contrast.

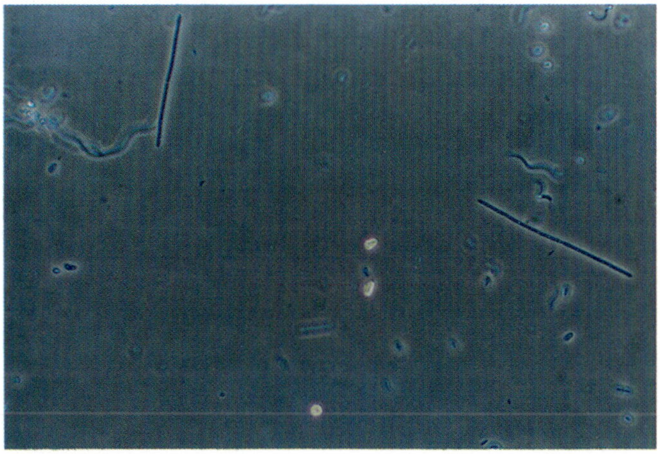

Figure 3. Two large rod-shaped bacteria dividing. Each single rod ≈30 μm long, from final clarifier, Cookeville, TN. Phase contrast.

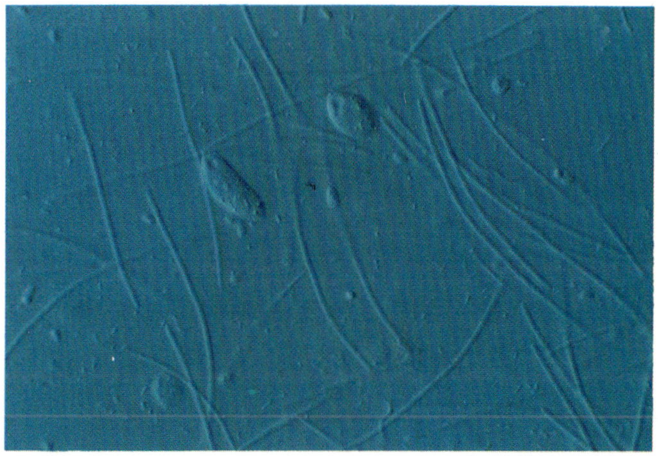

Figure 4. Many large rod-shaped and some small rod-shaped bacteria from final clarifier, Cookeville, TN. Differential interference contrast. Also in the photograph are several species of flagellates.

Figure 5. Spiral-shaped bacteria, possibly *Spirillum*, 20 to 25 μm long, from clarifier walls, Baxter, TN. Differential interference contrast. Other large filamentous bacteria and a filamentous green alga can be seen in the photograph.

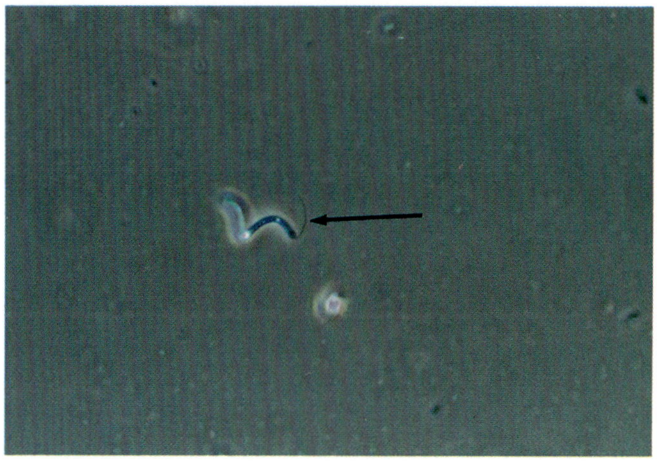

Figure 6. A spiral bacterium, ≈25 μm long, from activated sludge, Cookeville, TN. Phase contrast. The flagellum is visible at one end of the cell (arrow).

Figure 7. A large mass of various bacteria and flagellates from clarifier walls, Cookeville, TN. Differential interference contrast. Note several large filamentous bacteria (center and edge) that contain highly refractile granules. These are sulfur bacteria that deposit elemental sulfur within the cell.

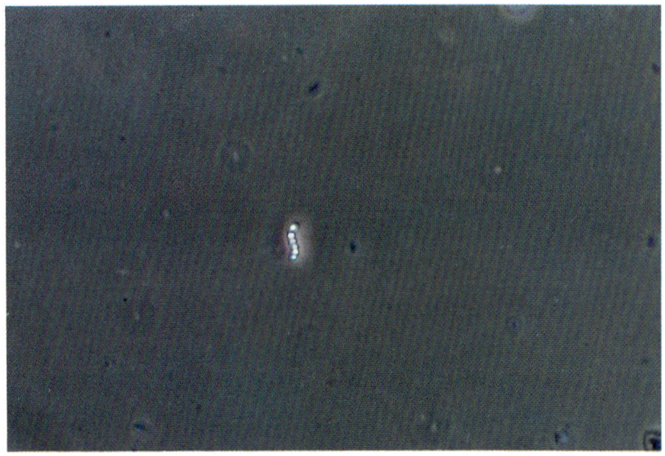

Figure 8. A bacterium with several highly refractile granules, ≈10 µm long, probably a sulfur bacterium, from activated sludge, Cookeville, TN. Phase contrast.

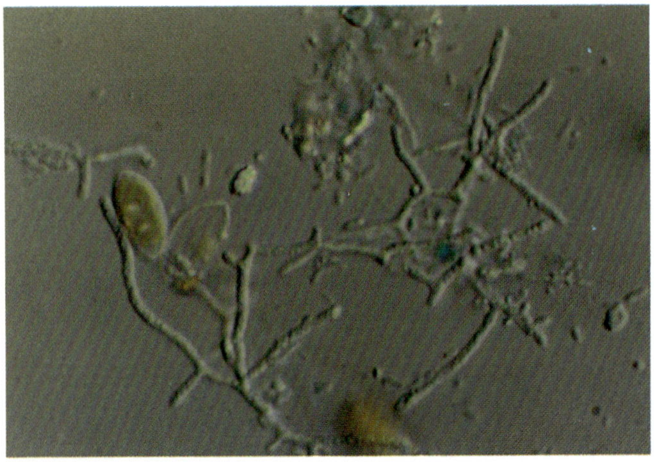

Figure 9. A cluster of bacteria that appears to be branching, forming a mycelium, ≈0.5 to 0.7 µm wide, from final clarifier, Cookeville, TN. Differential interference contrast. This is possibly *Nocardia*, which forms a primary mycelium and reproduces by fragmentation of the mycelium into rod-shaped or spherical nonmotile elements. Some actinomycetes have been mistakenly placed in the genus *Nocardia* based on morphology alone. Taxonomy is primarily based on cell envelope lipid and peptidoglycan composition. Some *Nocardia* species are pathogenic to man and animals. Population densities of *Nocardia* may be very high in foams of activated sludge systems.

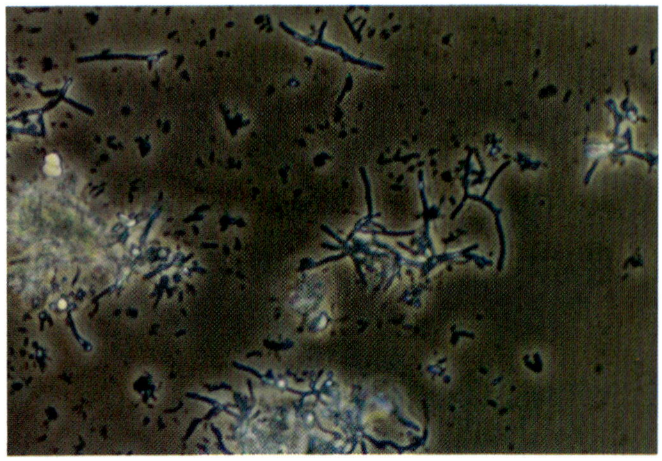

Figure 10. Another view of branching bacteria, *Nocardia* from final clarifier, Cookeville, TN. Phase contrast.

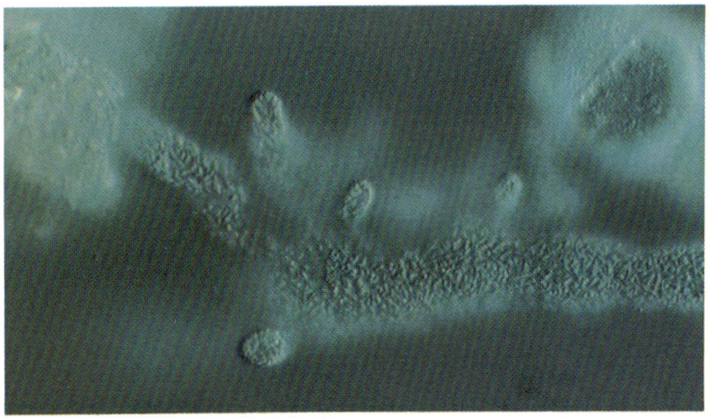

Figure 11. Cluster of bacteria with finger-like projections resembling a giant amoeba. These are *Zoogloea* from the bottom of the final clarifier, Cookeville, TN. Differential interference contrast. *Zoogloea* has only two species, both isolated from sewage or sewage treatment systems that utilize oxygen, such as activated sludge. The cells of *Zoogloea* are held near each other by a gelatinous matrix. The extracellular polymer is produced from conversion of dissolved organics in aerobic waste treatment.

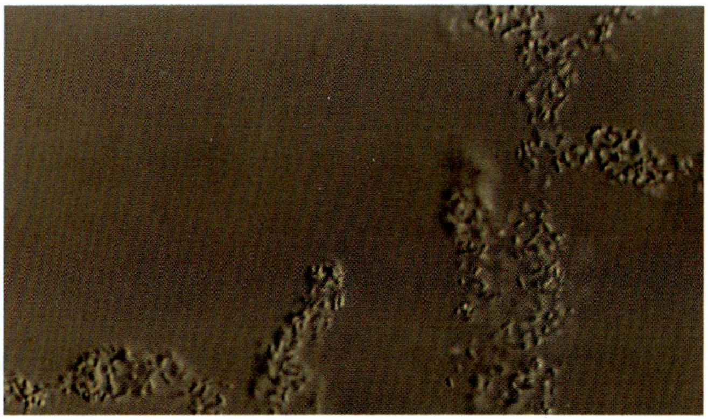

Figure 12. Another view of *Zoogloea* from activated sludge, Cookeville, TN. Differential interference contrast.

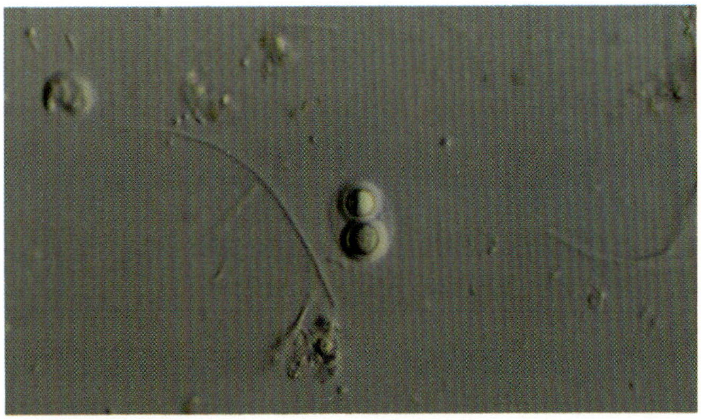

Figure 13. *Gloeocapsa*, 12 μm (together), from bottom of primary digester, Monterey, TN. Differential interference contrast. These are cyanobacteria surrounded by a thick sheath.

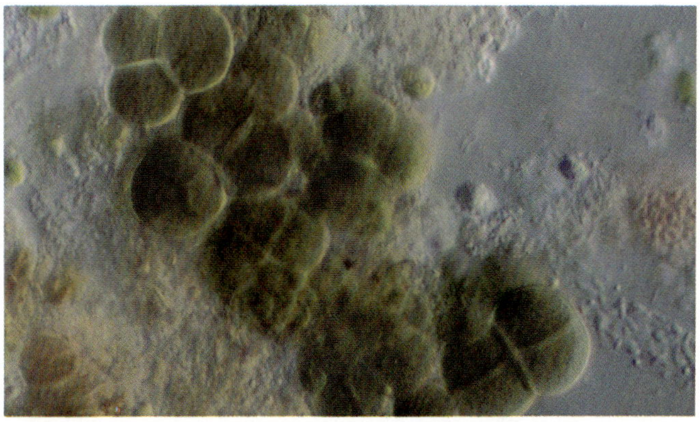

Figure 14. *Chroococcus*, 10 to 12 μm, from trickling filter material, Jefferson City, TN. Differential interference contrast. This is a cyanobacterium that forms aggregates of two to four cells due to failure of cells to separate immediately after division. They are surrounded by thin sheaths not visible in the photograph.

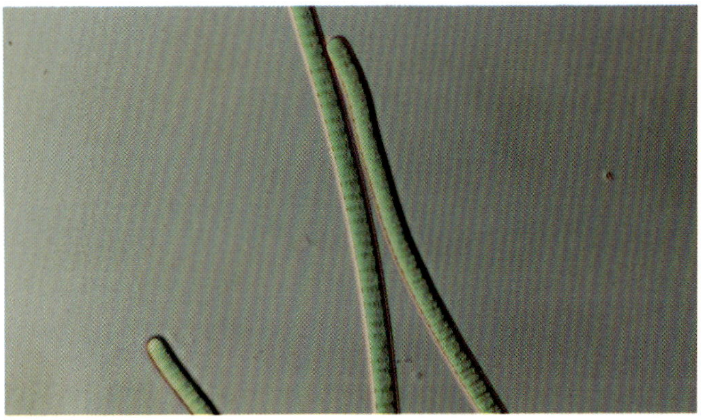

Figure 15. *Oscillatoria*, ≈10 μm wide, from walls of final clarifier, Cookeville, TN. Differential interference contrast. These are simple, large, unbranched filamentous cyanobacteria. They exhibit a gliding and oscillating motion.

Figure 16. *Lyngbya*, ≈10 μm wide, from trickling filter material, Jefferson City, TN. Differential interference contrast. This is a cyanobacterium with a distinct sheath seen in this photograph extending beyond the tip of the filament. Whether *Lyngbya* is a separate genus from *Oscillatoria* is controversial. Some argue that *Oscillatoria* produces distinctive sheaths under certain environmental conditions. Also in the photograph is a cluster of *Chroococcus* and some brown diatoms.

Flagellates

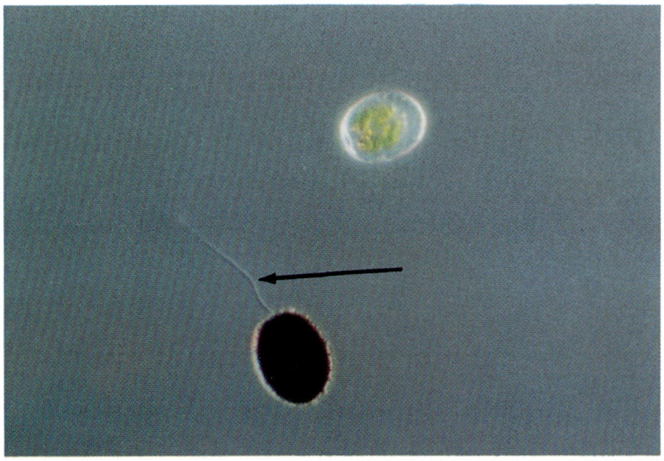

Figure 17. *Trachelomonas*. Differential interference contrast. This is a euglenoid that swims about in a lorica with a small opening at the anterior end through which the single flagellum (arrow) emerges. The lorica is impregnated with iron salts and becomes browner with age. Euglenoids may be photosynthetic or heterotrophic; *Trachelomonas* is photosynthetic, but the green color of the chloroplasts is frequently not visible through the lorica.

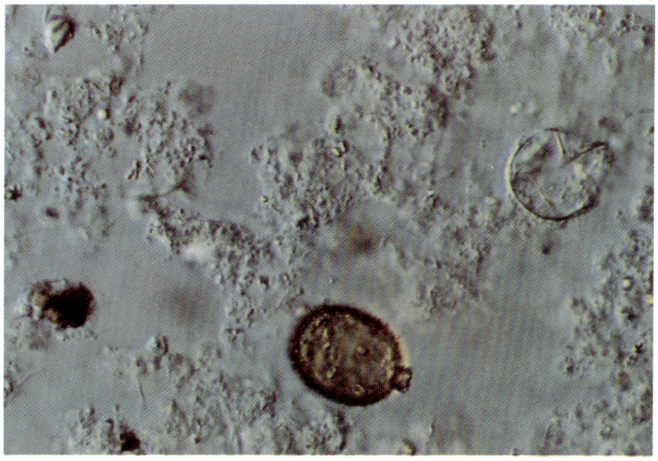

Figure 18. Empty lorica of *Trachelomonas* from activated sludge, Baxter, TN. Differential interference contrast. This photograph shows the small collar at the anterior end.

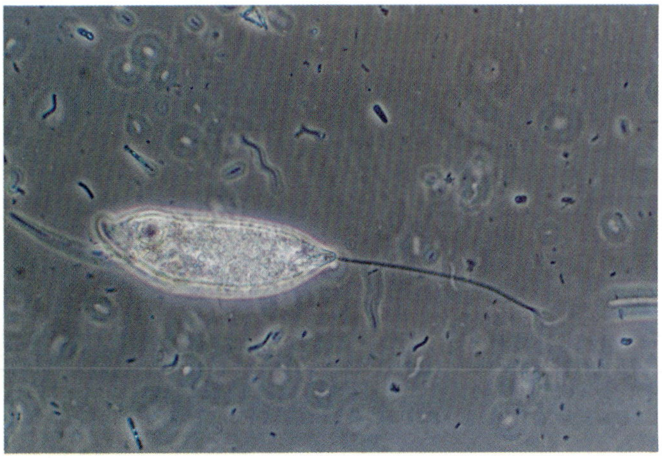

Figure 19. *Peranema*, 90 µm long (not including flagellum), from clarifier, Cookeville, TN. Phase contrast. This is a common, colorless, nonphotosynthetic euglenoid. It characteristically glides over surfaces with the anterior flagellum held stiffly straight ahead, with only the tip wiggling. *Peranema* is phagotrophic, capturing other organisms, including euglenoids, with a set of protrusible microtubular bundles located at the anterior end of the cell. A second flagellum runs backward tightly pressed to the body. It is very difficult to see. (In this photograph the filament extending from the back portion of the body is composed of filamentous bacteria; it is not the second flagellum.)

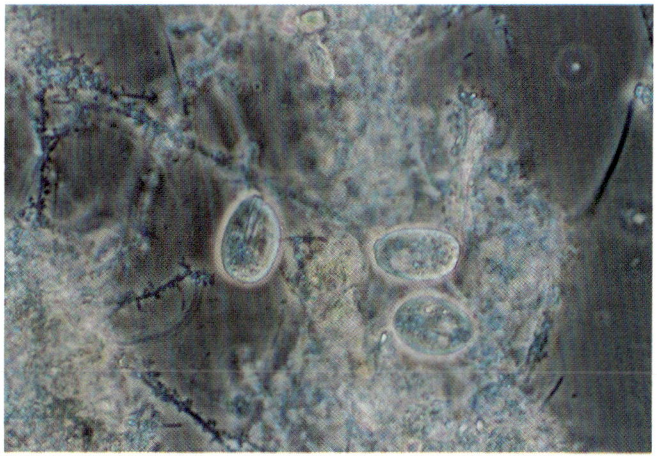

Figure 20

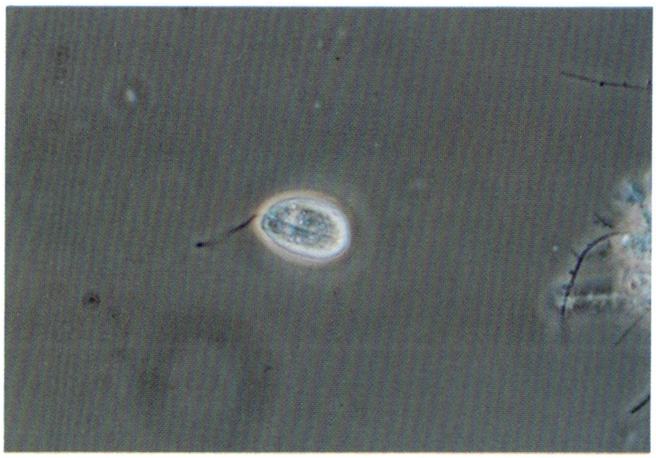

Figure 21

Figures 20, 21. *Entosiphon*, 20 µm long, from activated sludge, Cookeville, TN. Phase contrast. This is another phagotrophic euglenoid in which the ingestion apparatus is readily visible as a tube running nearly the length of the cell. It eats bacteria. Although it has two flagella, only the anterior-directed one is visible (**Figure 21**).

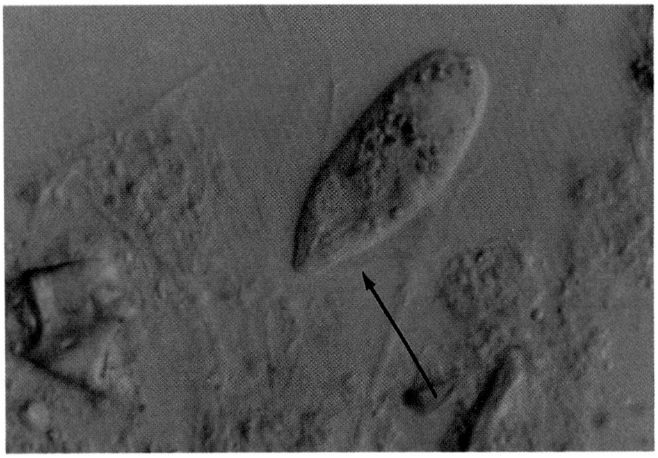

Figure 22

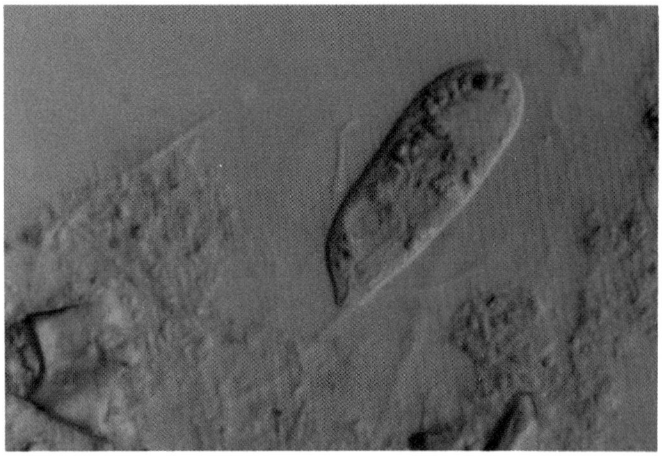

Figure 23

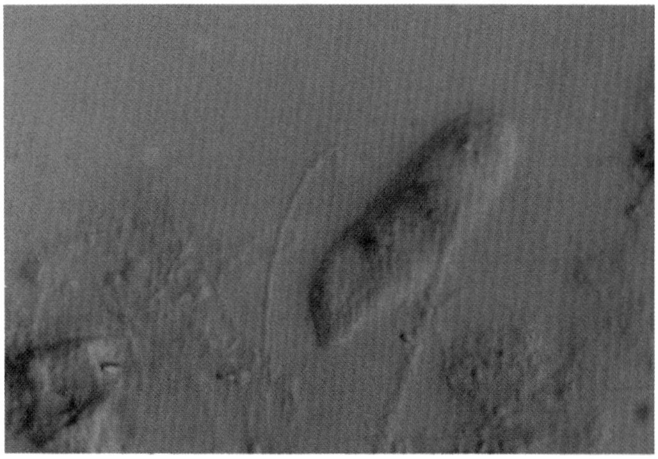

Figure 24

Figures 22, 23, 24. These are different optical sections through a colorless phagotrophic euglenoid flagellate, 35 μm long, from final clarifier effluent, Cookeville, TN. Differential interference contrast. The photographs show the rods of the ingestion apparatus at the anterior end (arrow, **Figure 22**) and the helically ridged surface of the cell (**Figure 24**) which is characteristic of euglenoids.

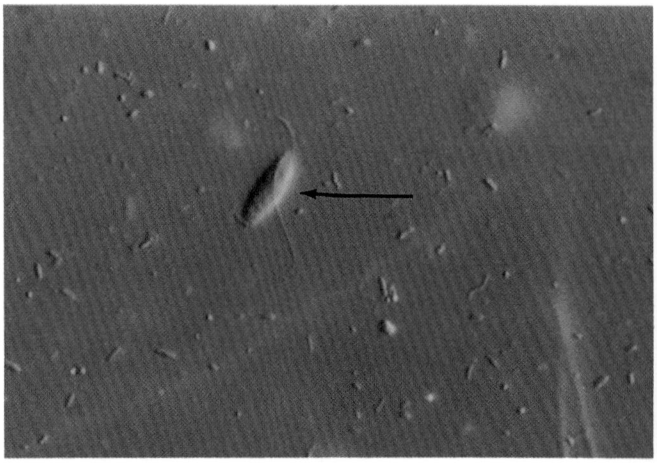

Figure 25

Figure 25. *Bodo* (arrow), 10 μm long, from trickling filter material, Jefferson City, TN. Differential interference contrast. Bodonids are related to euglenoids and have many similarities at the ultrastructural level. *Bodo* has two flagella that emerge at the anterior end of the cell. One is directed anteriorly, and the other posteriorly. These small, colorless flagellates ingest bacteria and also feed saprotrophically. Numerous species exist in wastewater.

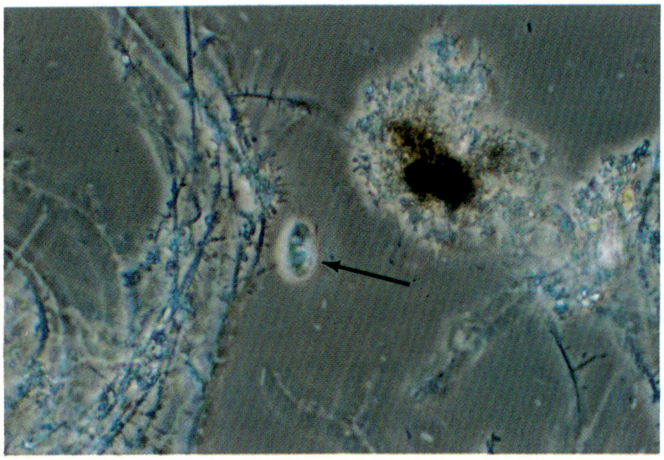

Figure 26

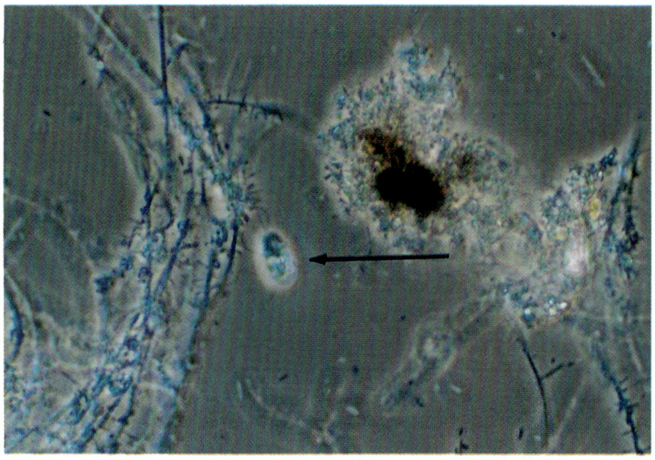

Figure 27

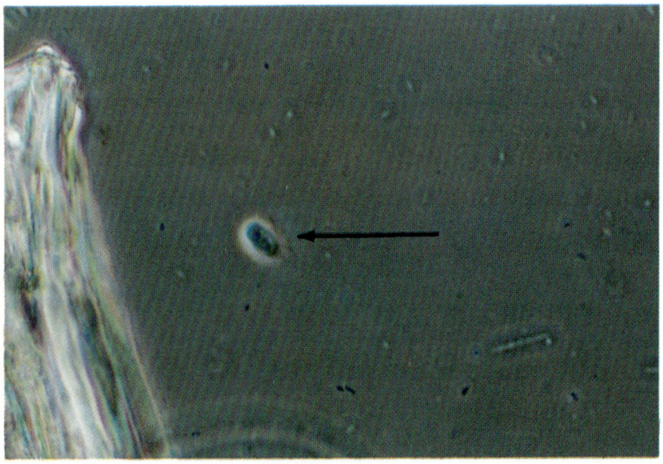

Figure 28

Figures 26, 27, 28. *Hexamita* (arrows), 12 μm, from activated sludge, Cookeville, TN. Phase contrast. *Hexamita* is a diplomonad, a group which contains mostly parasitic forms. (*Giardia* is a diplomonad.) Free-living diplomonads usually live in environments with little or no free oxygen. They eat bacteria or feed saprotrophically. Three genera of free-living diplomonads occur in wastewater: *Hexamita, Trigomonas,* and *Trepomonas*. All have two nuclei and eight flagella. *Trigomonas* is shaped like a triangle. *Hexamita* has posterior flagella extending beyond the body; *Trepomonas* does not.

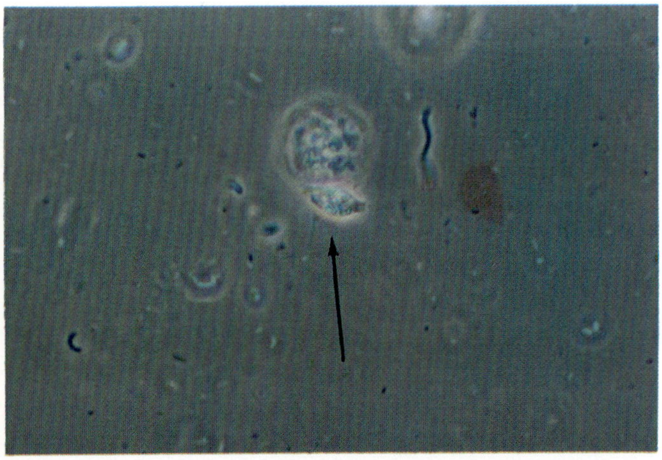

Figure 29. *Hexamita* (arrow), 12 μm, from clarifier, Cookeville, TN. Phase contrast. This is a different species from **Figures 26, 27, 28.**

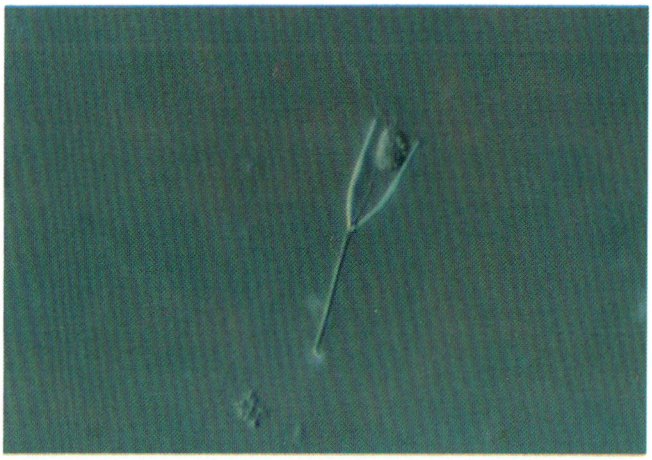

Figure 30. *Poteriodendron*, 18 μm long (cup), from activated sludge, Murfreesboro, TN. Differential interference contrast. The cell has two flagella, one of which extends backwards and attaches the cell to its lorica. The second draws a current of water towards the cell through its beating. Bacteria are trapped by the protruding flap of cytoplasm near the base of the anterior flagellum and ingested.

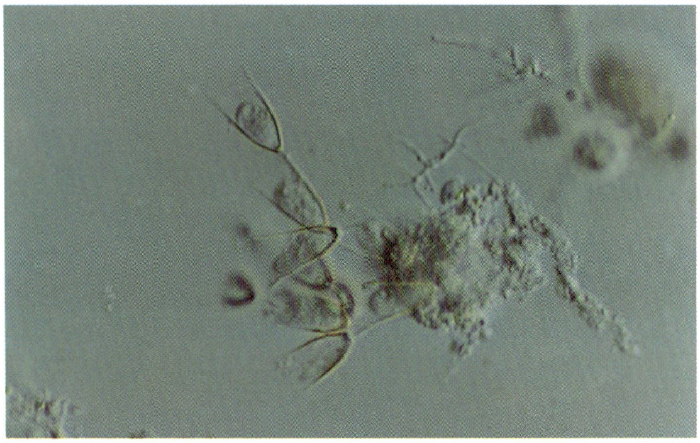

Figure 31. Colonies of *Poteriodendron,* from activated sludge, Murfreesboro, TN. Differential interference contrast. Colonies are built up as stalks of younger lorica are secreted in the cups of older cells.

Figure 32. *Histiona*, 10 μm wide, from final clarifier effluent, Cookeville, TN. Differential interference contrast. This organism feeds like *Poteriodendron.* The flap of cytoplasm is extended into a larger structure in this genus, and the lorica itself is much more delicate.

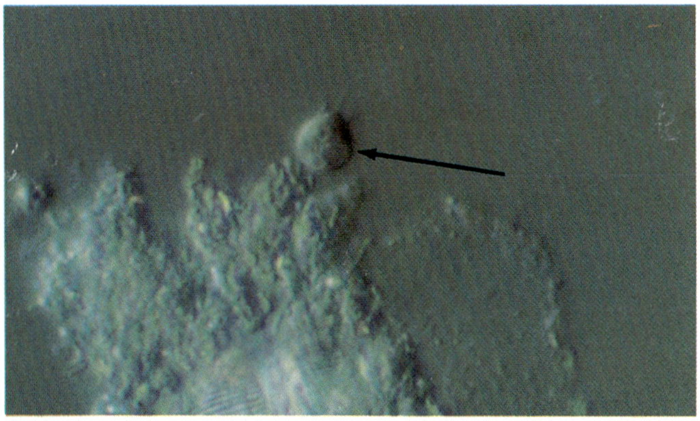

Figure 33. Choanoflagellate (arrow), 7 μm long, from activated sludge, Murfreesboro, TN. Differential interference contrast. These have a single flagellum at the anterior end of the cell. The beat of the flagellum draws water against a delicate collar that surrounds the base of the flagellum, and food is trapped there. The collar is visible as two little pegs in this photograph. Although the collar is circular, only one optical slice through it is in focus.

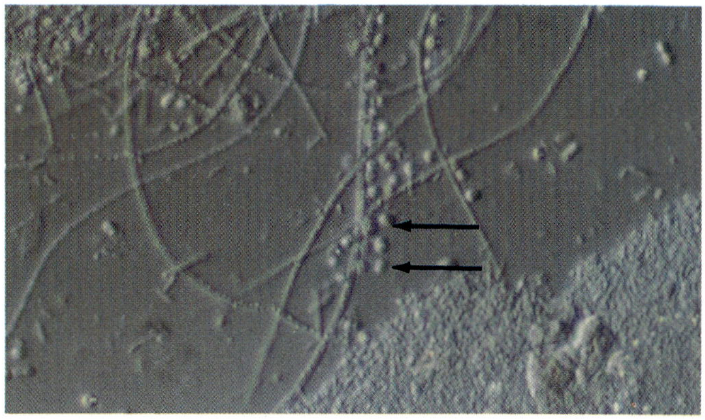

Figure 34. *Monas* (arrows), 2 to 3 μm in diameter, from trickling filter material, Jefferson City, TN. Differential interference contrast. *Monas* is a generic name that has been given to tiny, more or less spherical, colorless flagellates. An electron microscope would be required to determine the identity of flagellates in this genus. Many are probably quite unrelated to each other and do not belong in the same genus. This "*Monas*" is a tiny spherical form that lives in clumps attached to filamentous bacteria. It is attached by its posterior flagellum and bobs up and down as the flagellum contracts.

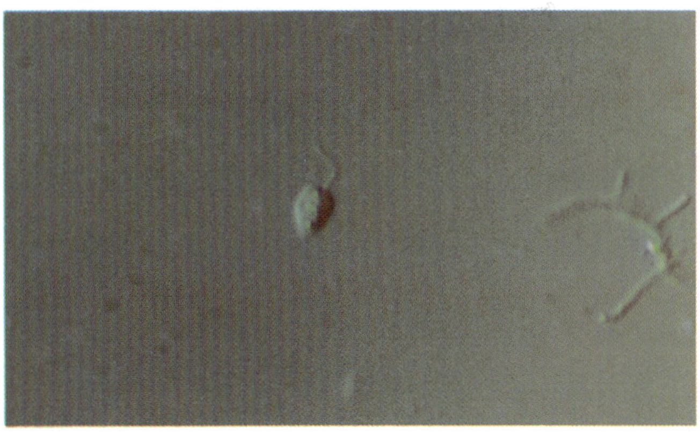

Figure 35. *Oicomonas* (or *Oikomonas*), 5 μm long excluding the flagellum, from activated sludge, Murfreesboro, TN. Differential interference contrast. *Oicomonas* is a name applied to many (probably unrelated) small, colorless flagellates with a notch at the anterior end near the emergence of one or two flagella. The body usually tapers posteriorly. The structure of some members of the genus indicates that they are related to golden-brown algae.

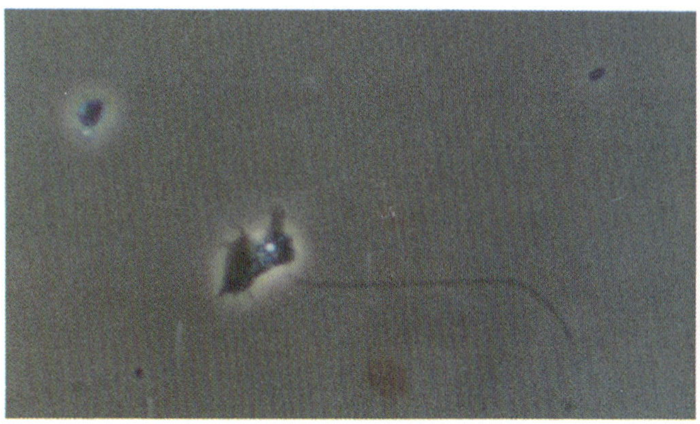

Figure 36. *Mastigamoeba*, 15 μm in longest dimension, from activated sludge, Cookeville, TN. Phase contrast. *Mastigamoeba* has both pseudopods and a flagellum; it combines the characteristics of amoebae and flagellates. *Mastigella* is a similar genus also found in treatment plants. In *Mastigamoeba*, the nucleus remains near the base of the flagellum, while there is no connection between the base of the flagellum and the nucleus in *Mastigella*.

Non-Flagellated Algae

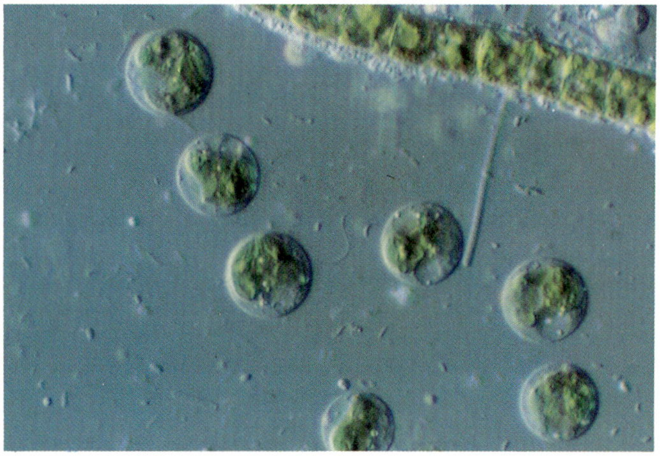

Figure 37. Unidentified round, green algae, 15 µm in diameter, from trickling filter material, Jefferson City, TN. Differential interference contrast. There are numerous genera of small, green spherical algae that can be distinguished by ultrastructure, life cycle studies, and pigment composition.

Figure 38. Filamentous alga of the family Cladophoraceae, probably *Rhizoclonium*, 50 to 60 µm wide, from clarifier walls, Baxter, TN. Differential interference contrast. These filaments may or may not be attached by a holdfast cell. They possess one or few nuclei compared with *Cladophora*, which may have many nuclei. They can be distinguished from *Oedogonium* by the absence of cell division rings (see **Figure 43**) in *Rhizoclonium*. Reproduction occurs by fragmentation or through production of small motile biflagellated zoospores.

Figure 39. Another specimen of *Rhizoclonium* from clarifier walls, Baxter, TN. Differential interference contrast.

Figure 40. Ends of two cells of *Rhizoclonium,* 70 µm at widest point. This is the same specimen as **Figure 39** but at a higher magnification.

Figure 41. *Cladophora*, 10 µm wide, from clarifier walls, Baxter, TN. Differential interference contrast. This is a branching species, cells of which are multinucleate. Reproduction may be sexual or asexual depending on species. In a few species, reproduction may be only by fragmentation.

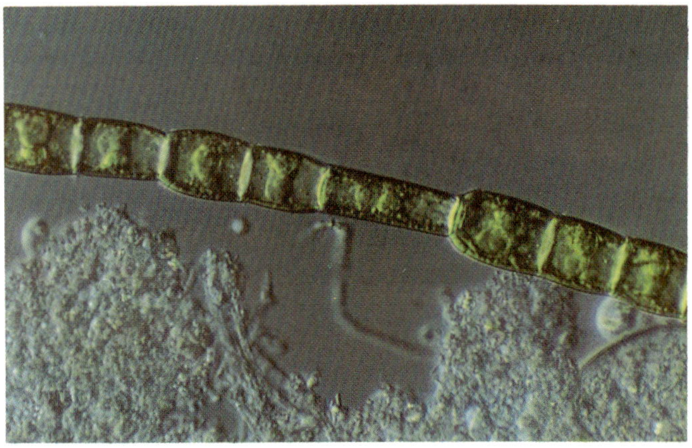

Figure 42. *Oedogonium*, 12 to 18 µm wide, from clarifier effluent, Murfreesboro, TN. Differential interference contrast. This genus contains several hundred freshwater species. This is an unbranched, filamentous alga. Cell division occurs in certain cells of the filament resulting in rings (**Figure 43**). Asexual reproduction occurs by fragmentation and production of flagellated zoospores. Sexual reproduction occurs through production of eggs and sperm. No sexual reproduction stages were observed in these specimens; however, germinating zoospores can be seen in **Figure 45**.

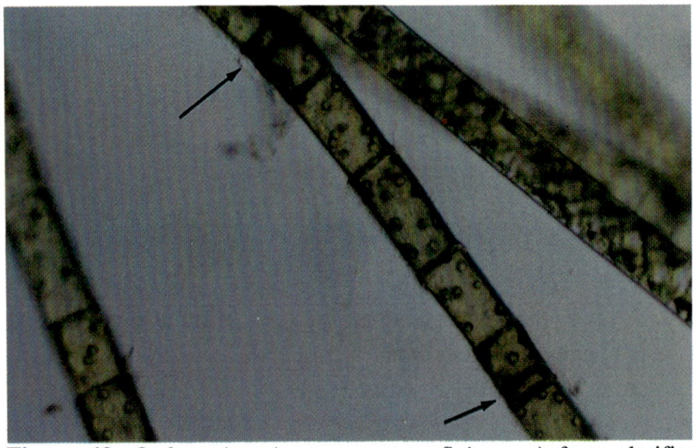

Figure 43. *Oedogonium* (center, next to *Spirogyra*) from clarifier walls, Cookeville, TN. Differential interference contrast. This specimen shows dividing cells (arrows) of *Oedogonium*. The *Spirogyra* filament contains cells having both clockwise and counterclockwise winding chloroplasts.

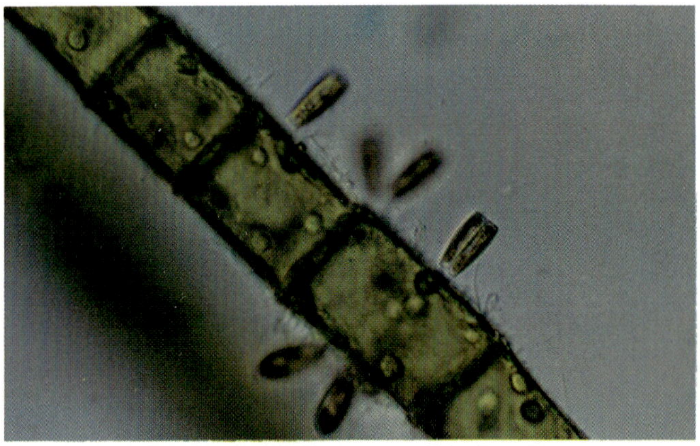

Figure 44. *Oedogonium* with diatoms, possibly *Gomphonema*, attached by one end to the side of the filament. The specimen was collected from clarifier walls, Cookeville, TN. Differential interference contrast.

Figure 45. *Oedogonium* with germinated zoospores attached to the side of the filament from clarifier walls, Cookeville, TN. Differential interference contrast. The zoospores may attach to any substrate.

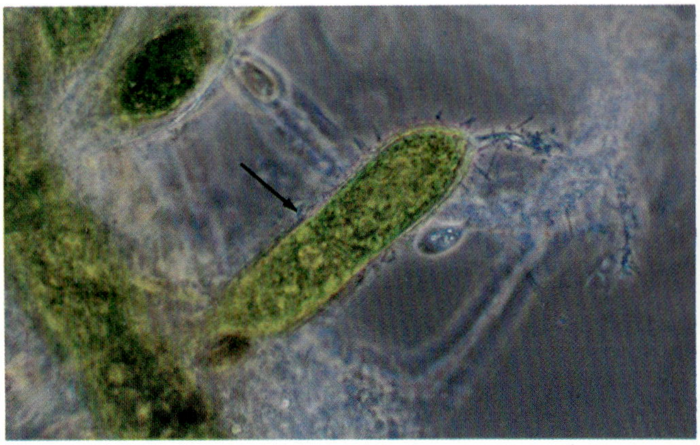

Figure 46. Germinating zoospore of *Oedogonium* from clarifier walls, Cookeville, TN. Phase contrast. The zoospore is forming a germling filament. Note the division of the new filament (arrow).

Figure 47. *Oedogonium* zoospore, 60 µm long, from clarifier walls, Cookeville, TN. Phase contrast. This specimen has large rod-shaped bacteria attached on the outside of the zoospore giving the appearance of cilia.

Figure 48. *Spirogyra*, 30 µm wide, from clarifier walls, Cookeville, TN. Differential interference contrast. This filamentous algal genus contains more than 275 species. Spiral band-like chloroplasts are scalloped at the margins (**Figure 49**) and contain numerous pyrenoids. Reproduction can be either asexual by fragmentation or sexual by means of conjugation between pairs of cells. A ciliated protozoan containing diatoms can also be seen in the photograph.

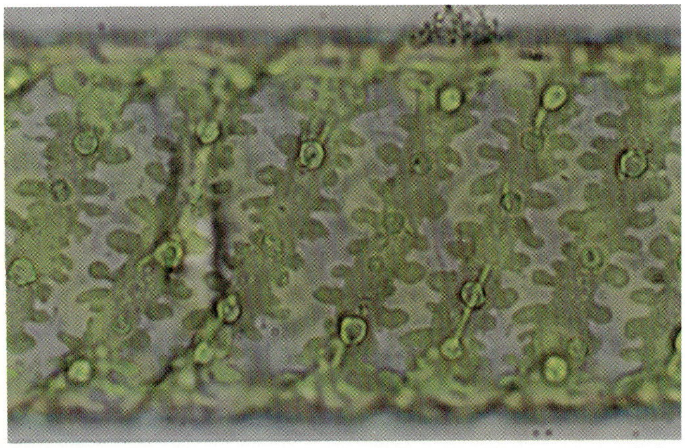

Figure 49. *Spirogyra*, 65 µm wide. This is a close-up view of the chloroplast. Photograph taken by Edie Sutherland using brightfield microscopy.

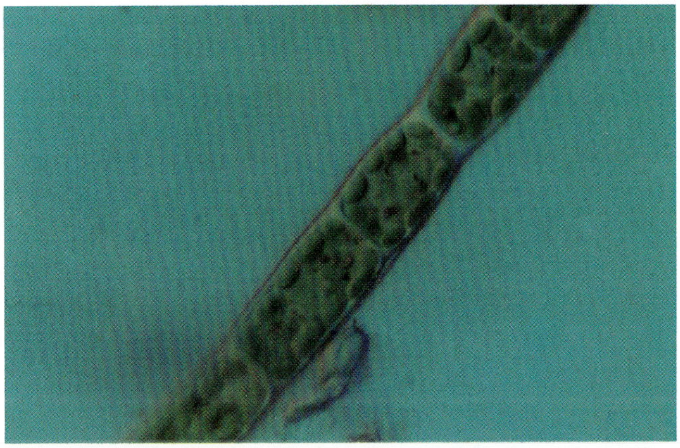

Figure 50. *Microspora*, 10 µm wide, from bottom of clarifier, Cookeville, TN. Differential interference contrast. Cell walls are thick and H-shaped. Chloroplasts lack pyrenoids. Reproduction can occur by fragmentation or production of flagellated zoospores. Union of isogamous, biflagellated gametes has also been reported.

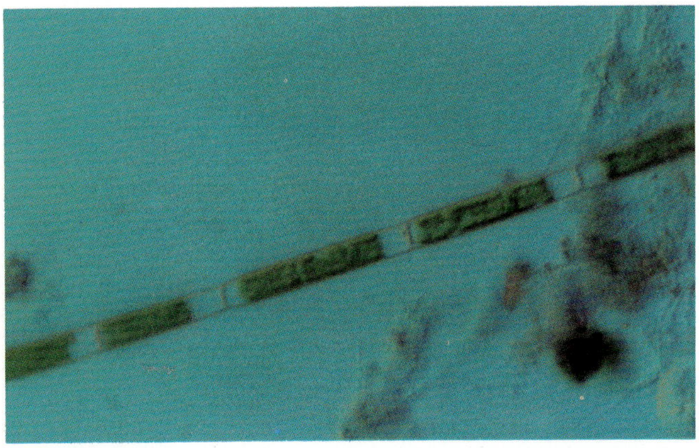

Figure 51. *Mougeotia*, 8 µm wide, from bottom of clarifier, Cookeville, TN. Differential interference contrast. This is an unbranched, filamentous alga with flat chloroplasts containing pyrenoids. The plastids may orient in different positions by rotating on their long axes. This specimen contains chloroplasts in face view.

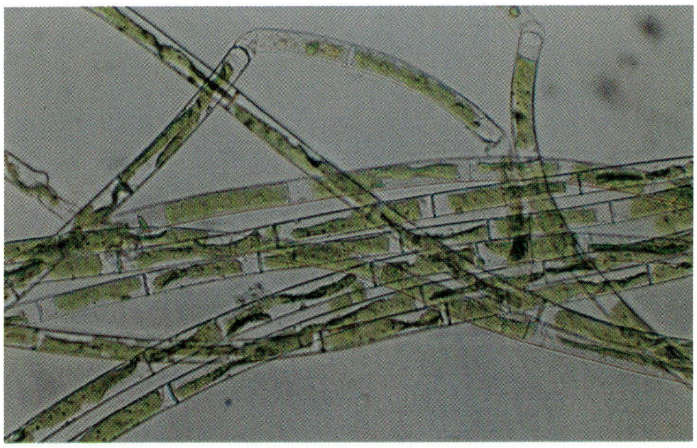

Figure 52. Another specimen of *Mougeotia* with chloroplasts oriented in various positions. Reproduction is by conjugation. Photograph taken by Cheryl Branson using brightfield microscopy.

Figure 53. *Stigeoclonium*, 7 to 10 μm wide, from final clarifier, Cookeville, TN. Brightfield. This is a widely distributed, branching filamentous alga. Asexual reproduction by zoospores has been reported although little is known about their life history.

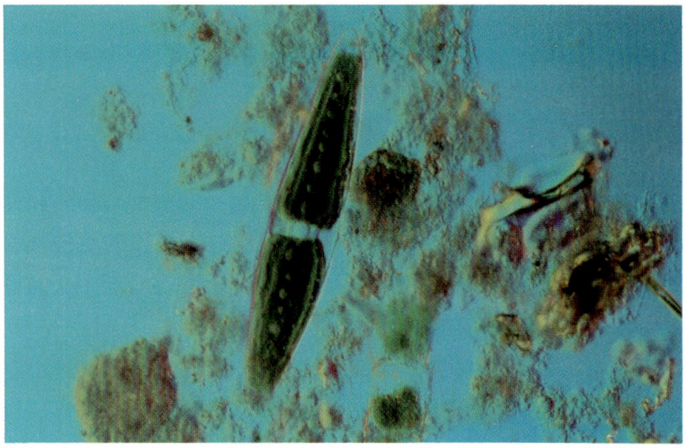

Figure 54. *Closterium*, 220 μm long, from bottom of final clarifier, Cookeville, TN. Differential interference contrast. This organism is a large unicellular desmid, narrowed toward both poles. Each half of the cell contains a chloroplast, and there is a large central nucleus between them. There are pores in the cell walls through which mucilaginous material is secreted. This mucilage helps it to move across a substrate. Reproduction is both sexual and asexual.

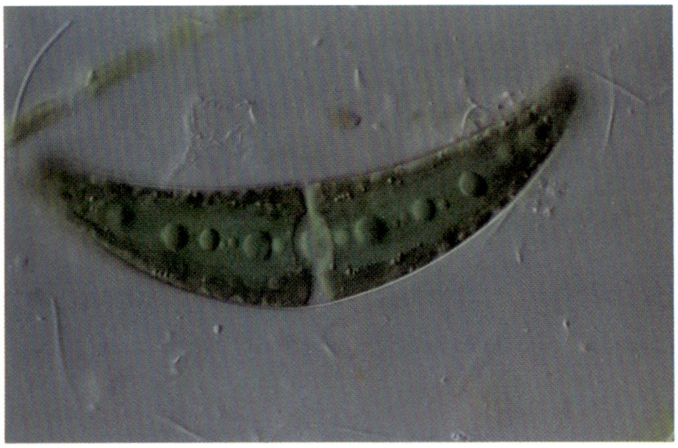

Figure 55. *Closterium*, 200 μm long, from final clarifier, Cookeville, TN. Differential interference contrast. This is another specimen as in **Figure 54**, but the cell is more curved.

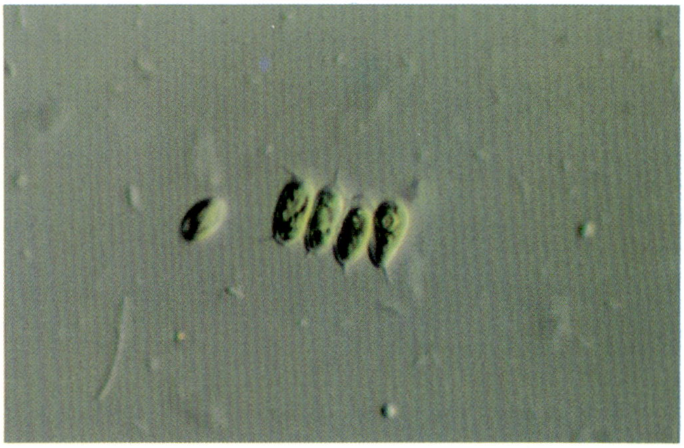

Figure 56. *Scenedesmus*, each cell ≈5 x 10 μm, from trickling filter material, Lincoln, NE. Differential interference contrast. Cylindrical cells of this genus are joined laterally in groups of four or eight (rarely 16). The terminal cells and, in some species, the others have spines. In this photograph, all four cells have spines.

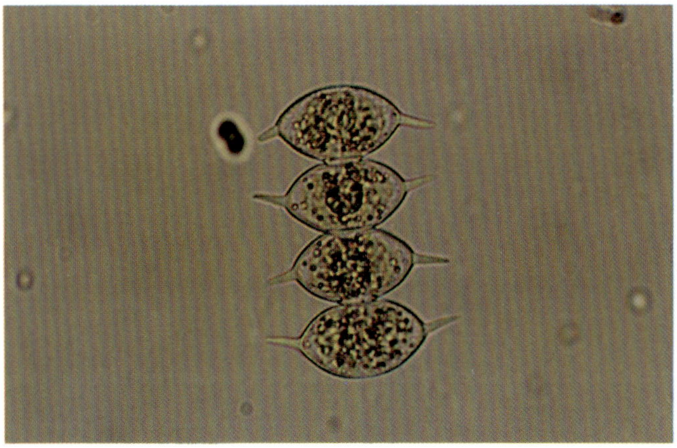

Figure 57. Another view of a *Scenedesmus* species, each cell ≈15 x 30 μm. Brightfield. Each cell has a spine. This may be more typical of how they would appear under brightfield microscopy. Photograph by Sujata Srikanth using brightfield microscopy.

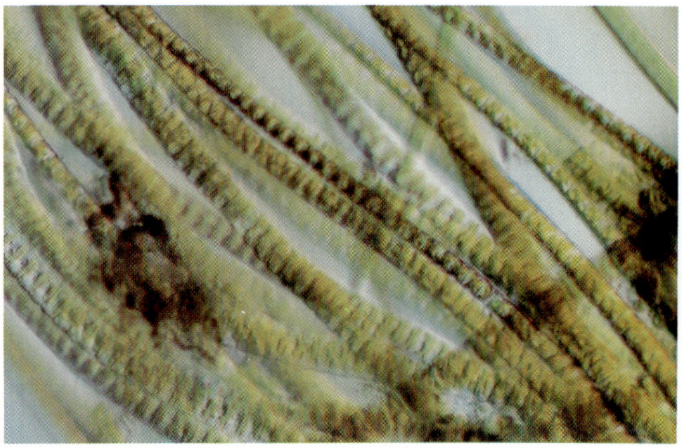

Figure 58. Filamentous concentric diatoms, possibly *Melosira* or *Coscinodiscus*, from clarifier walls, Cookeville, TN. These are diatoms that are classified almost entirely by the structure and ornamentation of the cell wall, the frustule. Such detail cannot be seen in the photograph.

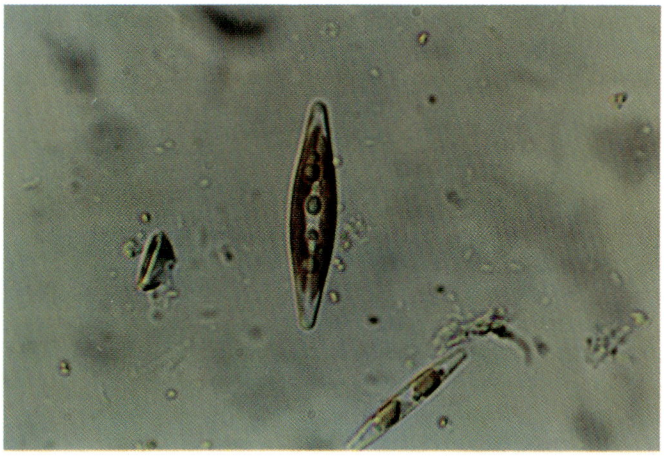

Figure 59. *Navicula* (center), 40 x 8 µm, and *Nitzschia* (lower portion of photograph), 30 x 4 µm, from activated sludge, Cookeville, TN. Differential interference contrast. Diatoms appear to glide along surfaces possibly on mucopolysaccharide material secreted by the cell.

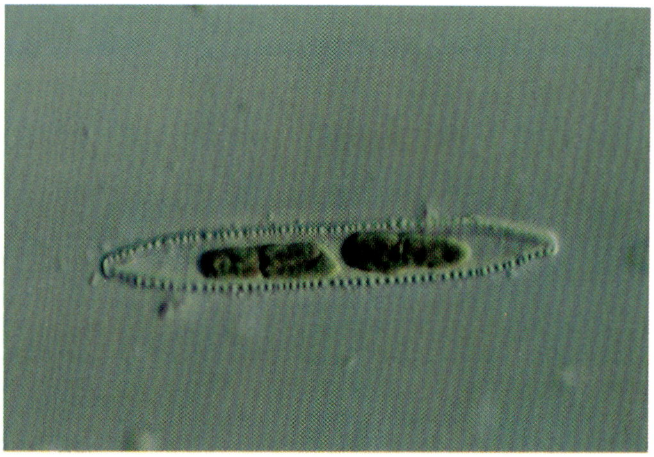

Figure 60. *Nitzschia*, 70 x 10 µm, from trickling filter material, Lincoln, NE. Differential interference contrast.

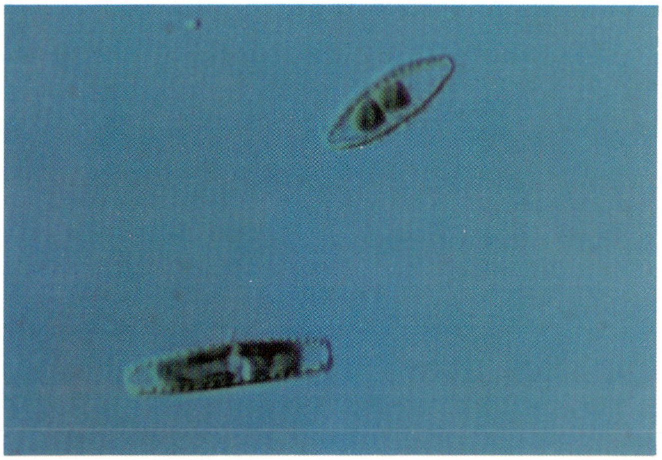

Figure 61. *Diatoma* in girdle view (blunt ends), 25 µm long, and in valve view (pointed ends), 20 µm long from clarifier walls, Murfreesboro, TN. Differential interference contrast.

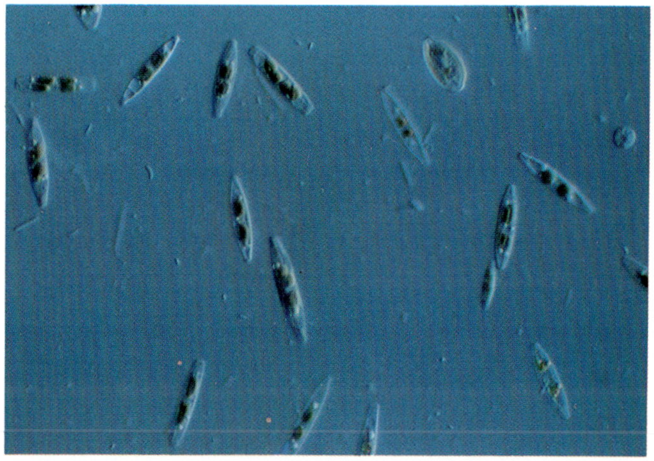

Figure 62. A group of diatoms, mostly *Nitzschia*, 25 to 30 µm long, from final clarifier, Cookeville, TN. Differential interference contrast.

Amoebae

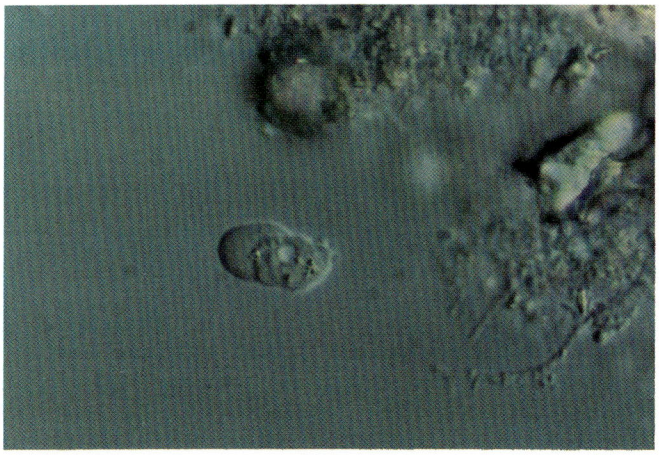

Figure 63

Figure 63. A vahlkampfiid amoeba, 18 μm, from the bottom of final clarifier, Cookeville, TN. Differential interference contrast. Vahlkampfiids can usually be recognized by the way they move. They form one pseudopod at a time, so that the body advances as one elongated blob. The advancing part of the pseudopod is clear and flows out rapidly, almost as though a small amount of liquid is poured out from the end of the cell. Three genera (*Naegleria, Tetramitus,* and *Vahlkampfia*) have been reported from wastewater. There is no reliable way to recognize to which of these genera an individual belongs without culturing them. Two genera produce a flagellated stage as well as an ameboid stage, and this can be useful for identification. *Vahlkampfia* does not produce flagellated stages; *Naegleria* produces flagellates (containing two flagella) that only survive for several hours before reverting back to amoebae; *Tetramitus* produces flagellates (containing four flagella) that feed and divide.

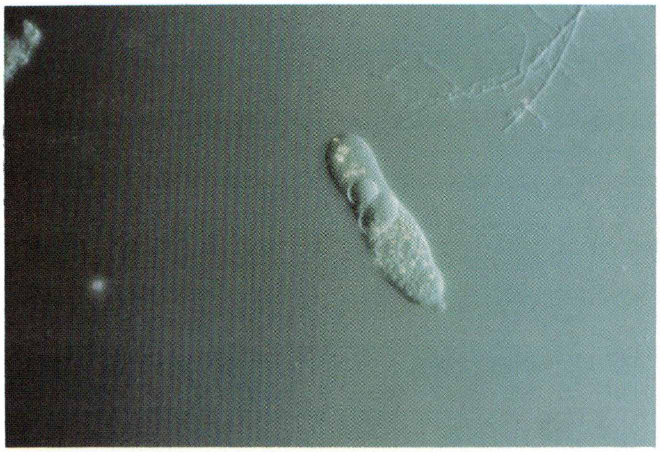

Figure 64

Figure 64. A hartmannellid amoeba, 100 μm, from bottom of final clarifier, Cookeville, TN. Differential interference contrast. Hartmannellids can usually be recognized without much difficulty. They are also monopodial (having one pseudopod). The clear area in the advancing part may be reduced, as compared with the vahlkampfiids, or absent.

Figures 65, 66. *Thecamoeba*, 100 μm, from clarifier scrapings, Murfreesboro, TN. Differential interference contrast. *Thecamoebae* are monopodial amoebae that have a large hyaline (clear) area running around the front and sides of the cell. Longitudinal ridges on the surface are present and extend in the direction of movement. They are sluggish, slow-moving amoebae that give the impression of having a thick, stiff surface.

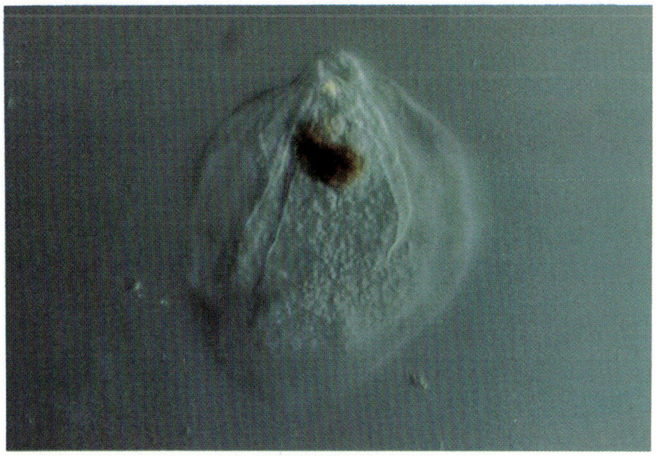

Figure 65

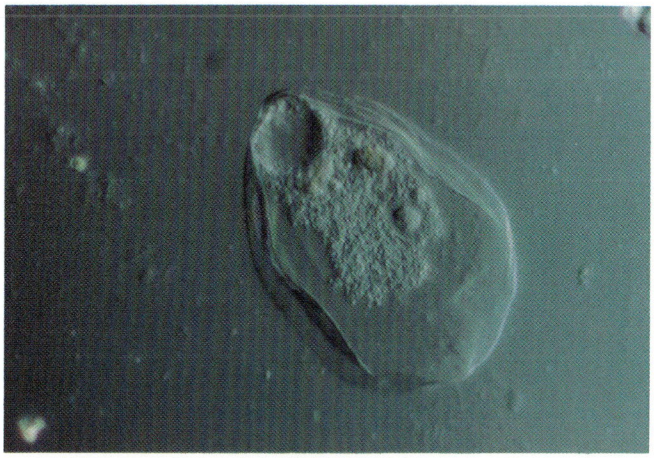

Figure 66

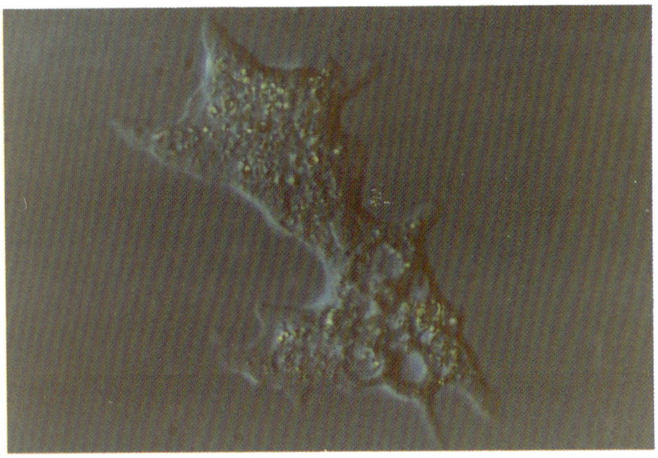

Figure 67. Mayorellid amoeba, 90 µm long, from aeration basin, Athens, TN. Differential interference contrast. Mayorellids form pseudopods that have conical or finger-like extrusions and form several at a time. There are numerous species of mayorellids, and they are very common.

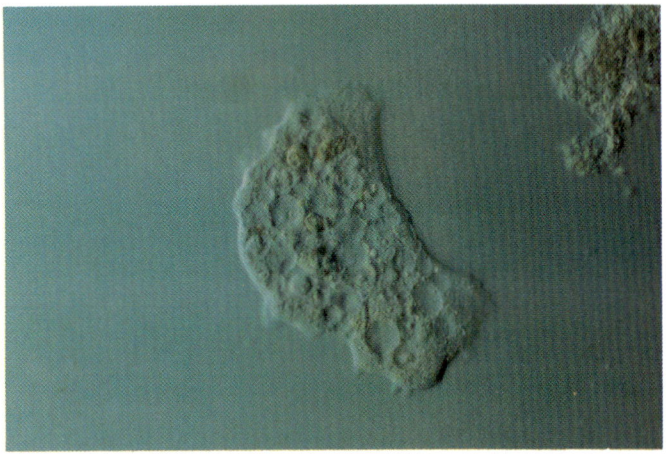

Figure 68. Unidentified amoeba, 140 µm long, from final clarifier, Cookeville, TN. Differential interference contrast. Many amoebae that cannot be readily assigned to a family or genus by sight occur in wastewater.

Figure 69. Unidentified amoeba, 150 µm, from final clarifier, Cookeville, TN. Differential interference contrast.

Figures 70, 71. *Arcella*, 60 µm test diameter, from aeration basin, Athens, TN. Differential interference contrast. This is a common testate amoeba. The shell or test is made of small proteinaceous blocks. Mineral salts can be incorporated into the test; the brown color is due to iron salts. A long pseudopod can be seen extending to one side. The cell moves by extending one pseudopod at a time, attaching the tip to the substrate and then contracting the pseudopod to drag the test forward. The outline of the test aperature, through which the pseudopods extend, can be seen in the center of the test.

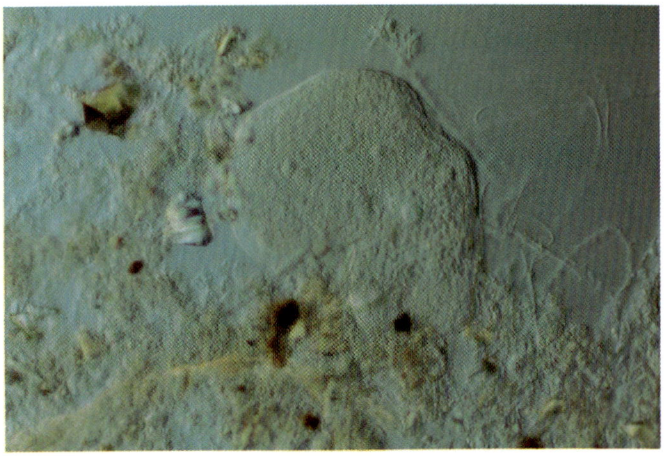

Figure 69

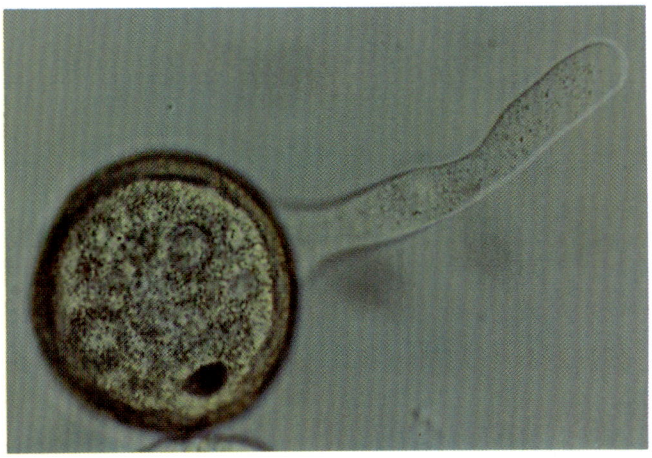

Figure 70

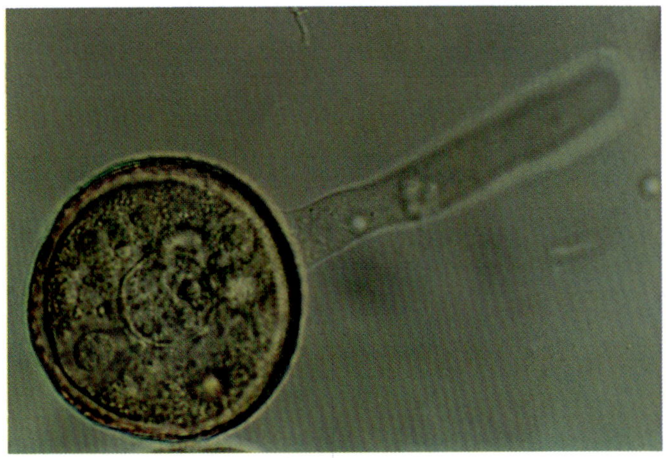

Figure 71

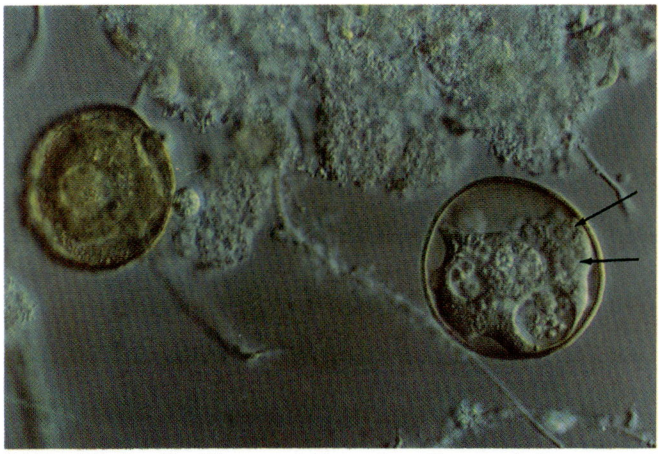

Figure 72. *Arcella*, 50 µm and 60 µm (test diameters), from secondary clarifier effluent, Murfreesboro, TN. Differential interference contrast. The cytoplasm is connected to the test by small pseudopods extending at irregular intervals from the cell. *Arcella* cells have two nuclei, which can be seen in the lobe of cytoplasm in one of the cells (arrow).

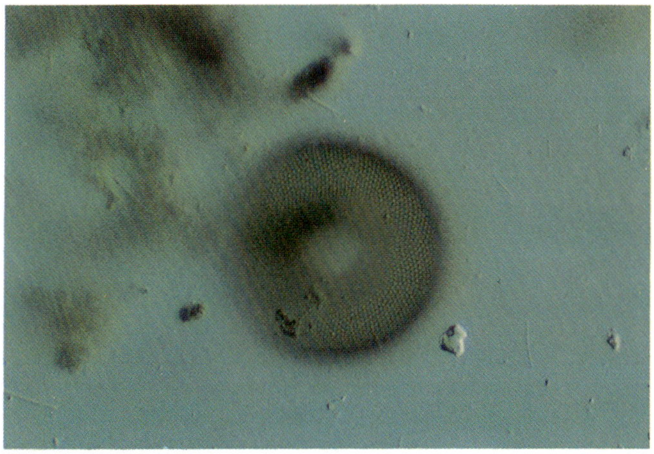

Figure 73. *Arcella* from Baxter, TN. Differential interference contrast. In this photograph of an empty test, the finely reticulate nature of the surface can be seen.

Figure 74. *Centropyxis*, 140 µm long, from Baxter, TN. Differential interference contrast. This is another testate amoeba. Several spines can be seen on one side.

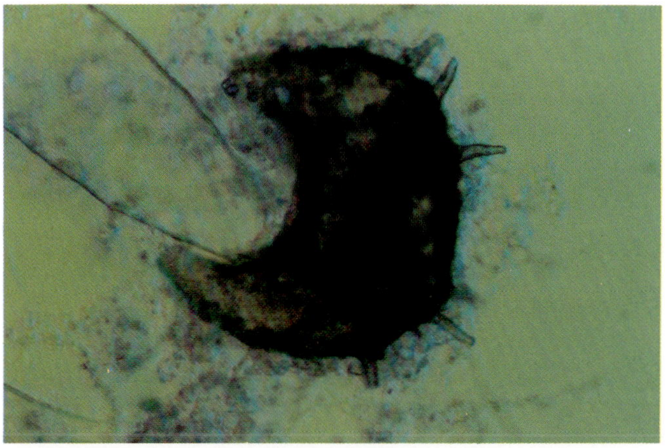

Figure 75. An empty amoeba test with five or six spines, 110 µm wide, from return sludge, Baxter, TN. Differential interference contrast.

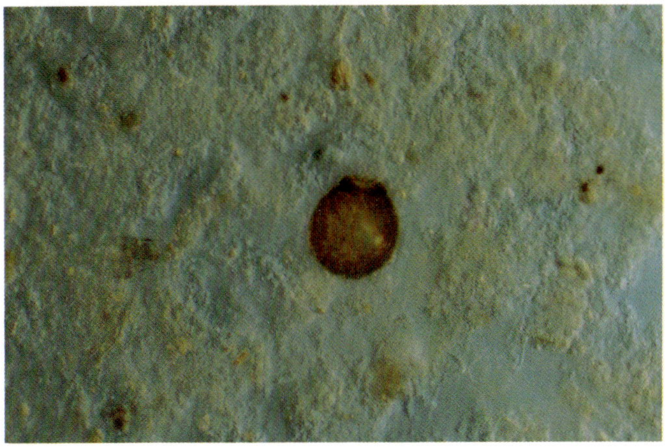

Figure 76

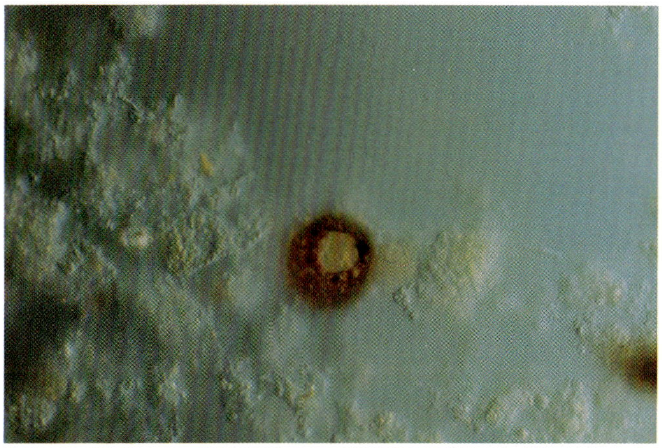

Figure 77

Figures 76, 77. *Difflugia* or *Pseudodifflugia*, test diameter ≈40 µm, from Baxter, TN. Differential interference contrast. Tests of both genera are made of agglutinated particles, which the amoebae collect from the substrate. *Difflugia* has lobose pseudopods like those of *Arcella*, while *Pseudodifflugia* has extremely fine, delicate pseudopods.

Figure 78. *Euglypha*, 70 μm (large individual) and 50 μm (small individual), from Baxter, TN. Differential interference contrast. These amoebae secrete small plates of silica, which they use to make their test. Before division occurs, the parent cell accumulates a large number of siliceous plates so that both amoebae will have complete tests after division. Accumulations of plates are visible at the aboral end of the cells.

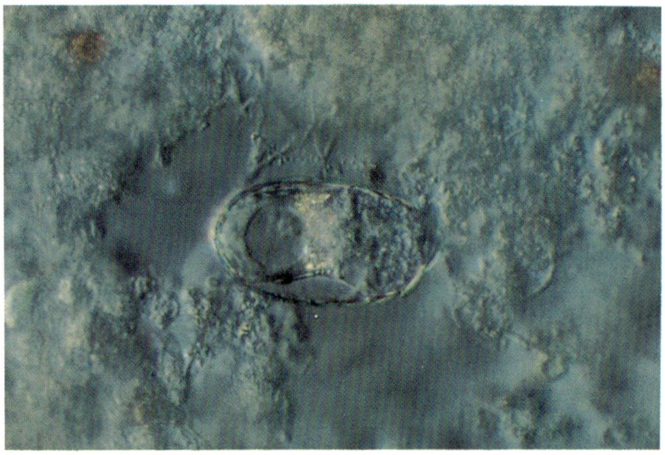

Figure 79. *Euglypha*, 50 μm long, from activated sludge, Murfreesboro, TN. Differential interference contrast. This is a live specimen showing the cytoplasm inside the test.

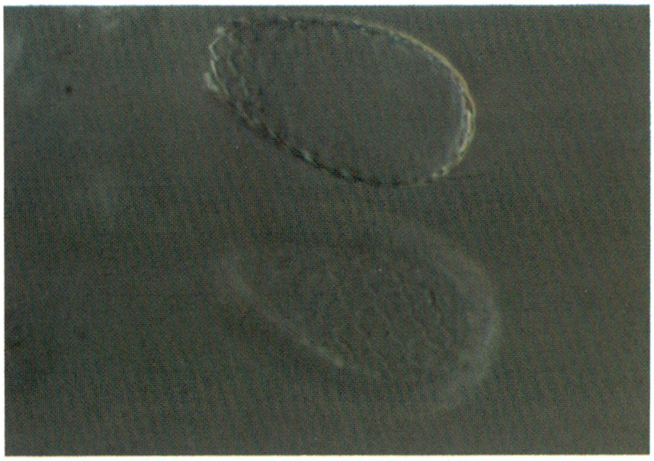

Figure 80. *Euglypha*, 60 μm long, from clarifier walls, Baxter, TN. Differential interference contrast. This is a double exposure of an empty test showing the arrangement of plates.

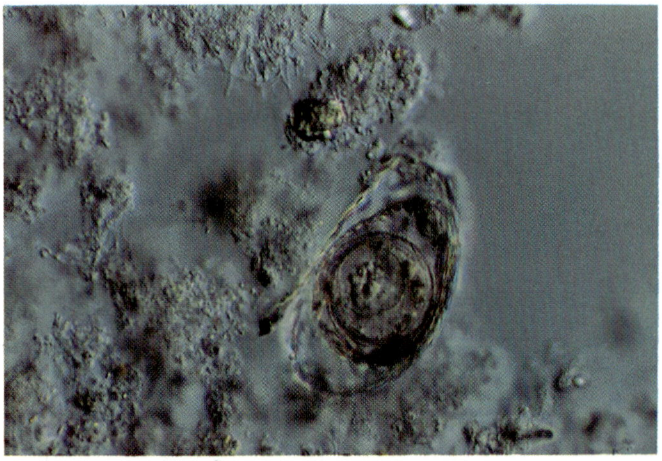

Figure 81. *Euglypha*, 70 μm long, from Baxter, TN. Differential interference contrast. *Euglypha* produces resistant cysts inside the old tests when conditions are unfavorable.

Ciliates

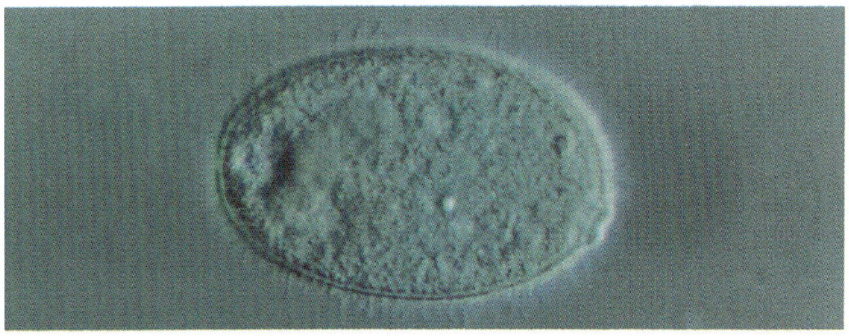

Figure 82. *Prorodon*, 70 μm long, from activated sludge, Watertown, TN. Differential interference contrast. It is a prostome ciliate that has an apical mouth (slightly shifted in the photograph due to coverslip pressure), uniform rows of cilia extending in meridians from pole to pole, and trichites (rods of microtubules) around the mouth. The contractile vacuole (water expulsion vesicle) can be seen at the posterior end of the cell, and just above it, the oval and slightly bent macronucleus.

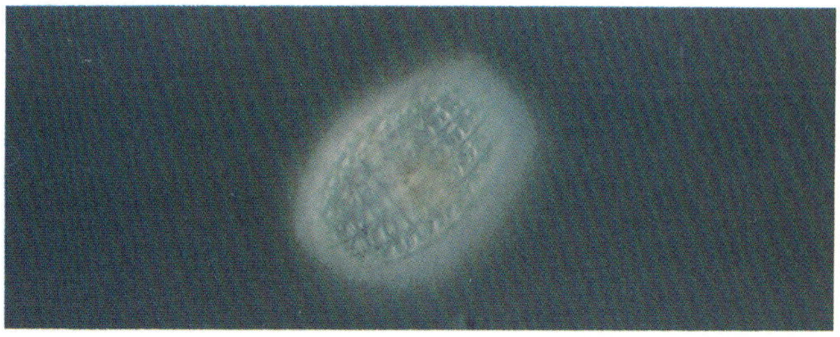

Figure 83. *Coleps*, 50 μm long, from the aeration basin, Murfreesboro, TN. Differential interference contrast. This ciliate is also a prostome, but the cellular organization is obscured by a coat of calcium carbonate plates, which cover the cell. This ciliate is a scavenger and eats tissue of other organisms including other ciliates.

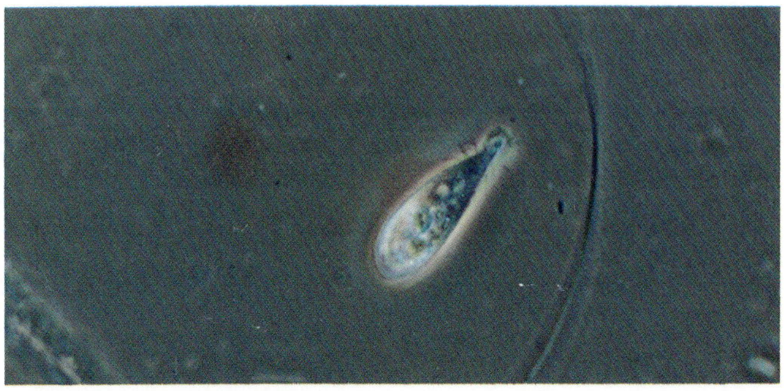

Figure 84. *Trachelophyllum*, 35 μm long, from activated sludge, Cookeville, TN. Phase contrast.

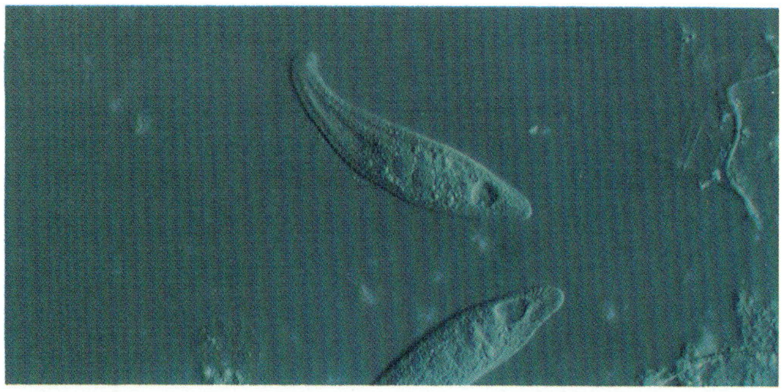

Figure 85. *Litonotus*, a pleurostome ciliate, 80 μm long, from activated sludge, Murfreesboro, TN. Differential interference contrast. The mouth is a slit at the anterior end of the cell (on the convex side of the curved part). It is laterally flattened and swims on its right side. Pleurostomes are carnivorous ciliates that eat flagellates and other ciliates.

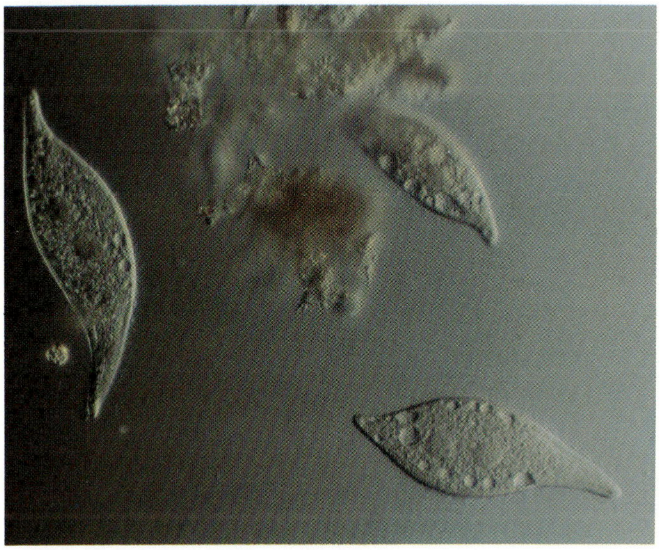

Figure 86. *Amphileptus*, 100 μm long, from activated sludge, Cookeville, TN. Differential interference contrast. This is another pleurostome genus with the same habits and appearance as *Litonotus*; the differences between them are trivial. In *Litonotus*, the ciliary rows along the body run parallel for the whole length, while they converge along a middle line in *Amphileptus*. *Litonotus* usually has a single water expulsion vesicle, while *Amphileptus* may have many. In this species, the water expulsion vesicles can be seen along the margins of the cell.

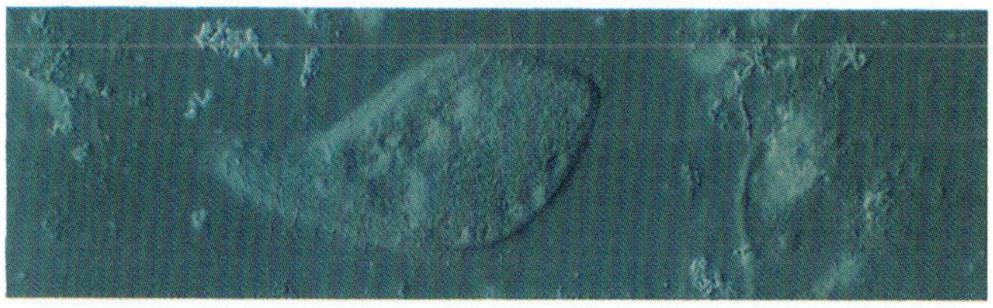

Figure 87. Another species of *Amphileptus*, 100 μm long, from activated sludge, Murfreesboro, TN. Differential interference contrast.

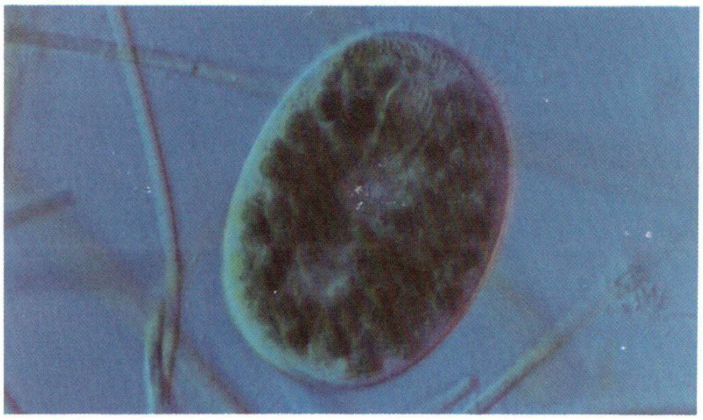

Figure 88. *Nassula*, 70 μm long, from clarifier scrappings, Murfreesboro, TN. Differential interference contrast. This organism uses a basket of microtubular rods (visible near the anterior end of the cell) to ingest algae, usually filamentous algae.

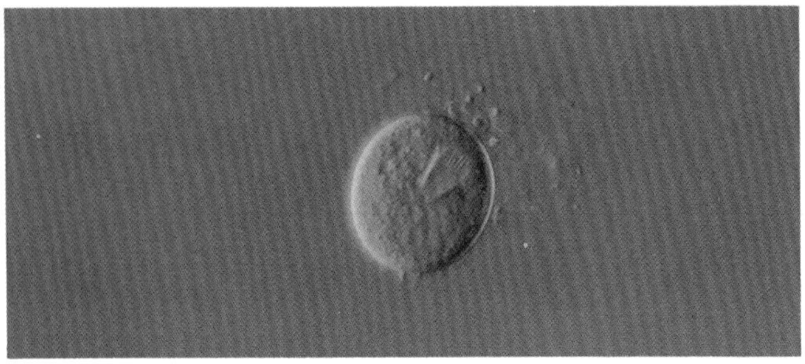

Figure 89. Encysted *Nassula*, 30 μm, from clarifier scrappings, Murfreesboro, TN. Differential interference contrast. The food vacuoles have disappeared, and the microtubular basket can be clearly seen.

Figure 90. A microthoracid ciliate, 40 μm, from activated sludge, Murfreesboro, TN. Differential interference contrast. Members of this group have very reduced ciliation and a thick stiff pellicle, giving them an armored appearance.

Figures 91–97. Various cyrtophorid ciliates of the genera *Chilodonella* and *Trithigmostoma.* All were photographed with differential interference contrast. These forms are very flattened with cilia only on the ventral surface of the body. They ingest bacteria and algae through a cytopharyngeal apparatus comparable to that of *Nassula. Chilodonella* has a large unciliated patch behind the mouth, whereas *Trithigmostoma* has cilia in the middle of the ventral surface; members of these two genera are the same in other respects.

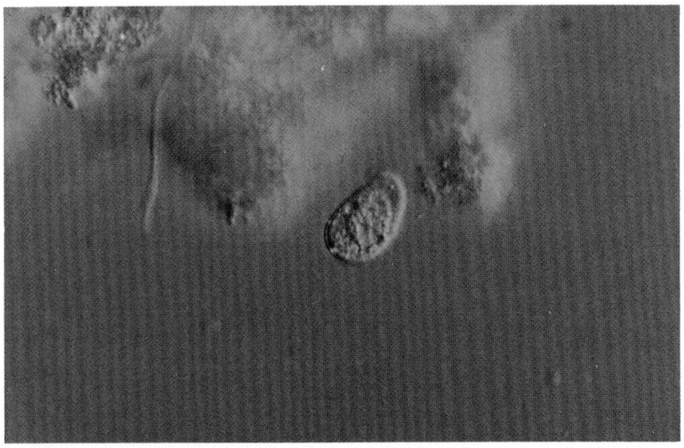

Figure 91. A cyrtophorid ciliate, 15 μm long, from activated sludge, Murfreesboro, TN.

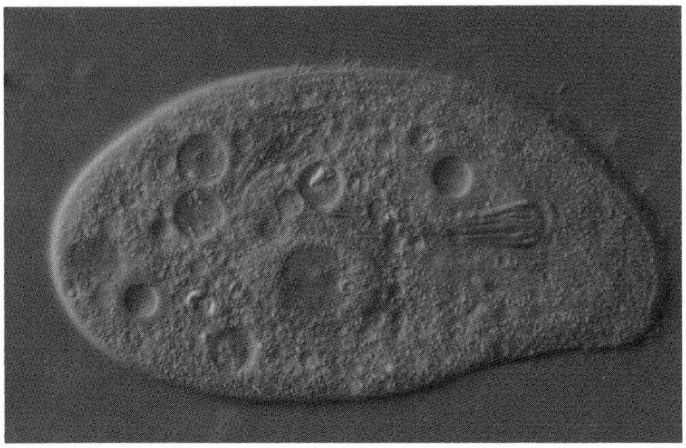

Figure 92. A cyrtophorid ciliate 140 μm long, from aeration basin, Murfreesboro, TN.

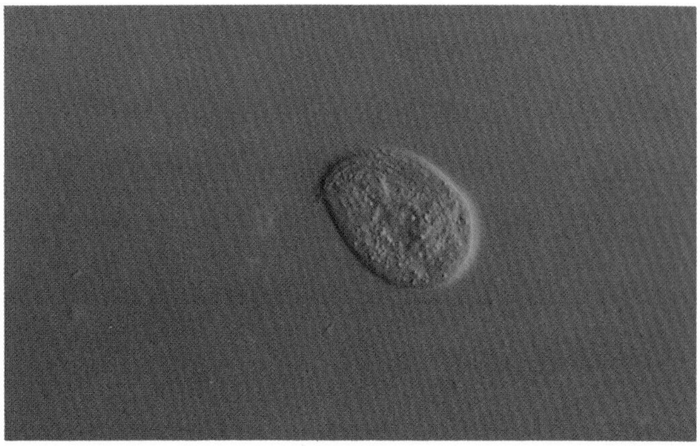

Figure 93

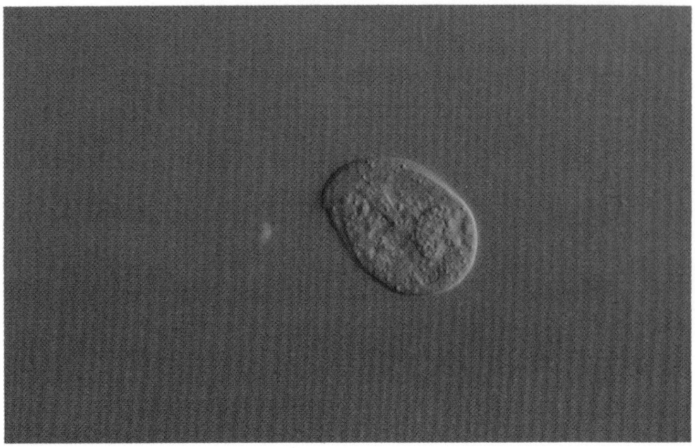

Figure 94

Figures 93, 94. The same cyrtophorid ciliate observed in various focal planes, 35 μm long, from aeration basin, Murfreesboro, TN.

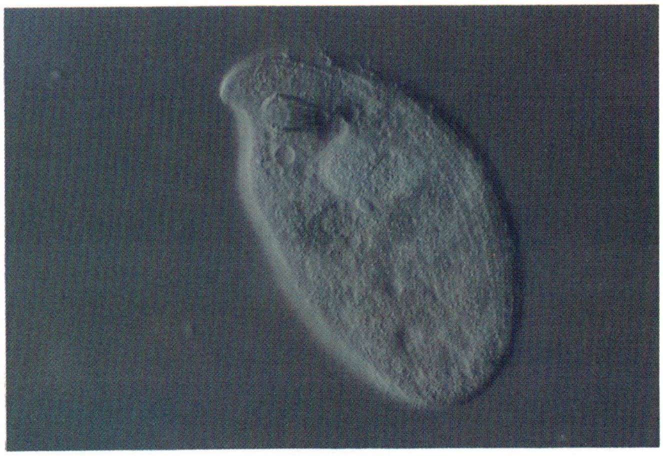

Figure 95

Figures 95, 96, 97. A cyrtophorid ciliate, 110 µm long, seen in various focal planes, from aeration basin, Murfreesboro, TN. Differential interference contrast.

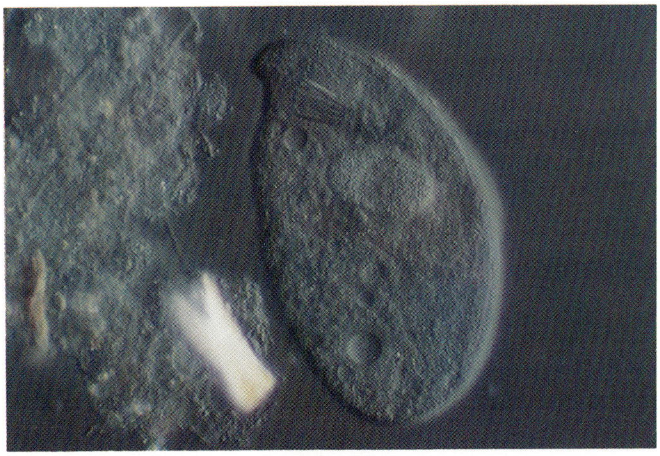

Figure 96

Figure 97

Figure 98

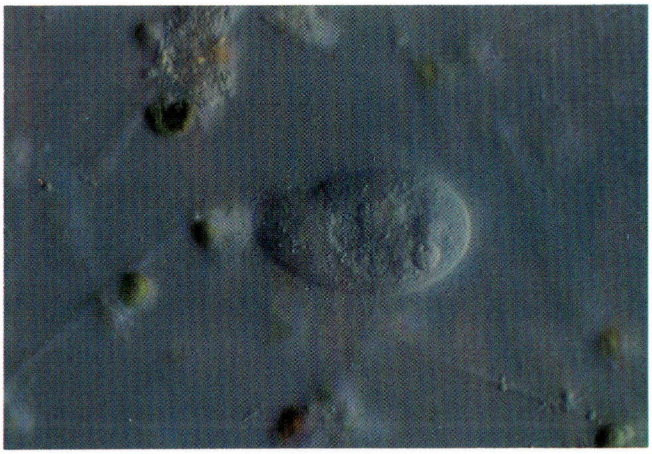

Figure 99

Figures 98, 99. *Colpidium*, a hymenostome ("membrane-mouth") ciliate, 40 µm long, from trickling filter material, Jefferson City, TN. Differential interference contrast. In addition to the uniform rows of cilia which cover the body, there are specialized clumps of cilia on either side of the mouth used to filter bacteria from the water. In *Colpidium*, these are very inconspicuous structures. **Figure 98** shows the ciliary rows on the surface of the body. **Figure 99** shows food vacuoles inside the cell. The top food vacuole is being formed as bacteria are filtered into it; a channel running from the food vacuole to the surface of the cell is visible.

Figure 100. *Paramecium*, 250 µm long, clarifier, Cookeville, TN. Differential interference contrast. This is another ciliate that filters bacteria from the water using specialized ciliary structures near the mouth. The shape of the cell is distorted by coverslip pressure, but many internal structures can be seen. The oral cilia can be seen in a cornucopia-shaped indentation in the middle of the cell. Numerous food vacuoles are present in the cytoplasm and contain bacteria in various stages of digestion. *Paramecium* has trichocysts that it discharges when attacked by predatory ciliates; they can be seen from above as little white dots below the mouth and from the side as rods lined up perpendicularly to the surface all around the cell. Cilia can be seen extending from the surface.

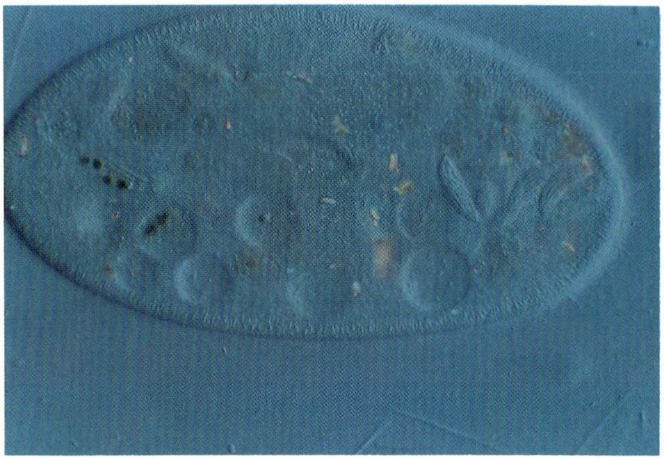

Figure 100

Figures 101, 102, 103. Water expulsion vesicles of *Paramecium* from final clarifier, Cookeville, TN. Differential interference contrast. Freshwater protozoa have water expulsion vesicles to get rid of excess water drawn into the cell by osmosis. Most species of *Paramecium* have two water expulsion vesicles, one at each end of the cell. Canals radiating out from a central vacuole fill up with water (**Figure 101**). They empty their contents into the spherical vacuole that they surround, disappearing as they do so (**Figures 102, 103**). The vacuole then empties its contents through a pore to the outside of the cell, and the radial canals begin filling with water again (bottom, **Figure 101**).

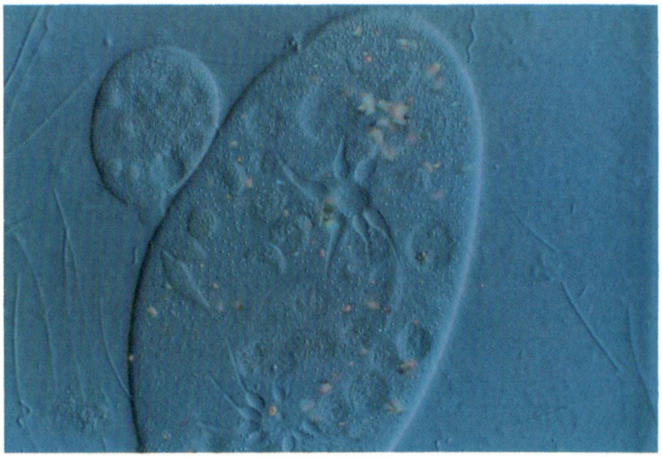

Figure 101

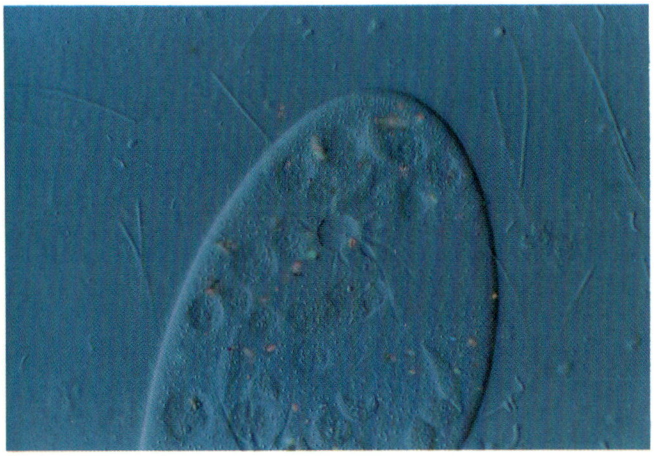

Figure 102

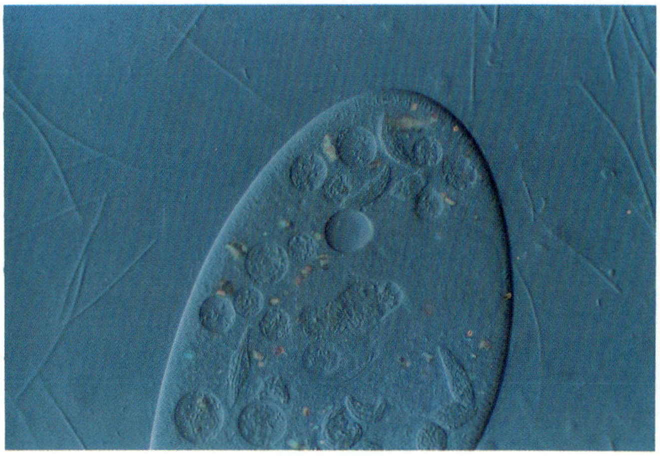

Figure 103

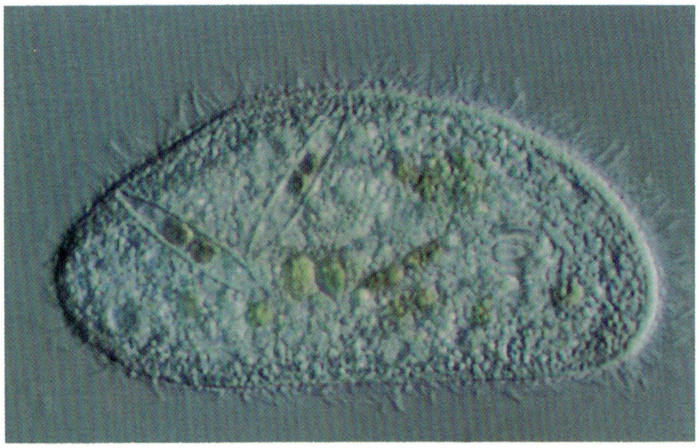

Figure 104. Unidentified ciliate, 85 μm long, from final clarifier, Cookeville, TN. Differential interference contrast. This ciliate has eaten diatoms, visible within the cell.

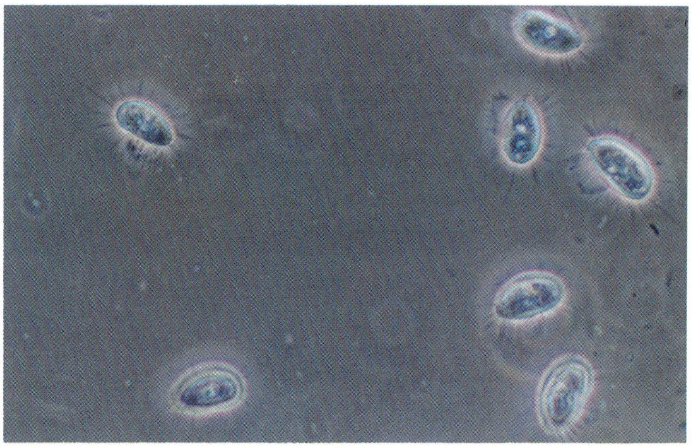

Figure 105. *Cyclidium*, 15 to 20 μm long, from trickling filter, Monterey, TN. Phase contrast. This is a scuticociliate that also filters bacteria with its oral cilia. An undulating membrane composed of a row of closely spaced cilia can be seen extending from the body like a small sail. The mouth is located near the back end of the undulating membrane. They also have a caudal cilium, longer than the rest, that extends from the posterior like a tail.

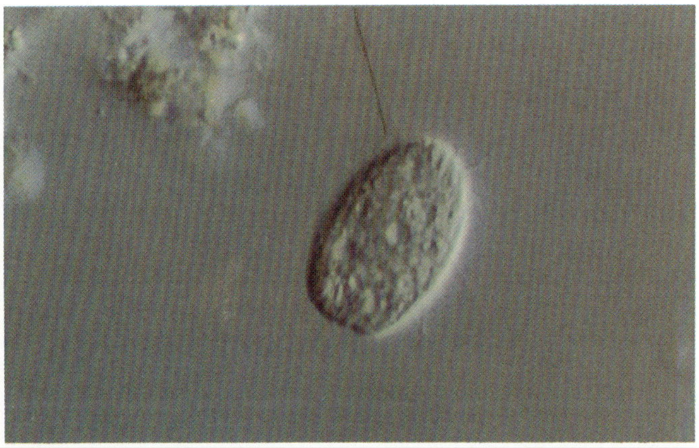

Figure 106. *Cinetochilum*, 30 μm long, from activated sludge, Cookeville, TN. Differential interference contrast. This is a another scuticociliate. The oral cilia are not as apparent as in *Cyclidium*, although they have the same general arrangement.

Figure 107. *Calyptotricha*, a scuticociliate 30 μm long, from bottom of final clarifier, Cookeville, TN. Differential interference contrast. This is a ciliate that lives in a very delicate lorica that it secretes. The lorica is attached to inanimate objects in the water. The oral cilia are well developed, but are not in the plane of focus in the photograph.

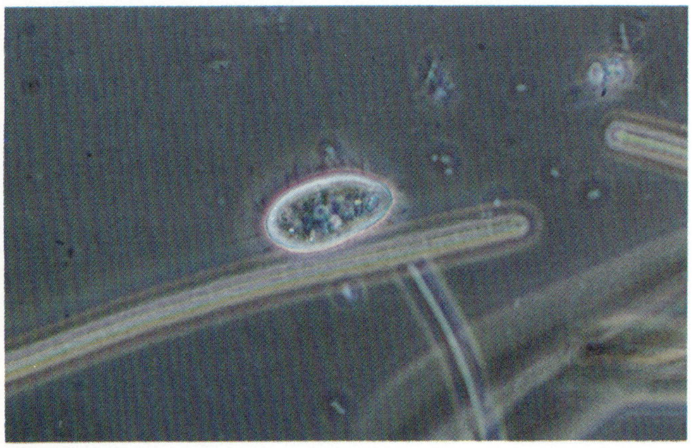

Figure 108. Unidentified scuticociliate, 30 μm long, from trickling filter, Monterey, TN. Phase contrast.

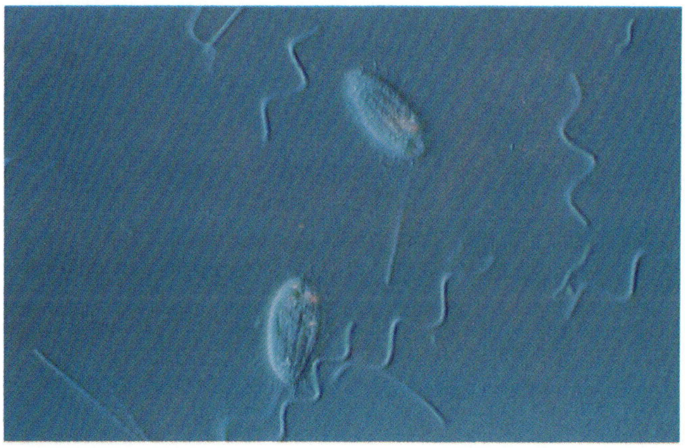

Figure 109. Unidentified scuticociliate, 30 μm long, with the bacteria *Spirillum*, from Clarifier, Cookeville, TN. Differential interference contrast.

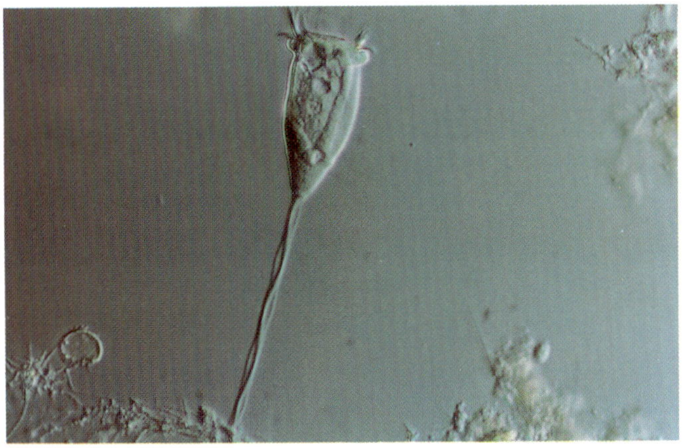

Figure 110. *Vorticella*, 170 µm in total length, from aeration basin, Athens, TN. Differential interference contrast. This is a peritrich ciliate. In peritrichs (peri= around, trich= hair), the oral cilia wind completely around the top of the cell. The stalk contains a contractile, slightly sinuous filament that can rapidly coil up like a spring, pulling the cell body down to the substrate. Peritrichs are filter-feeders that consume bacteria. They are abundant in aerobic environments containing a lot of bacteria, such as trickling filters and activated sludge basins.

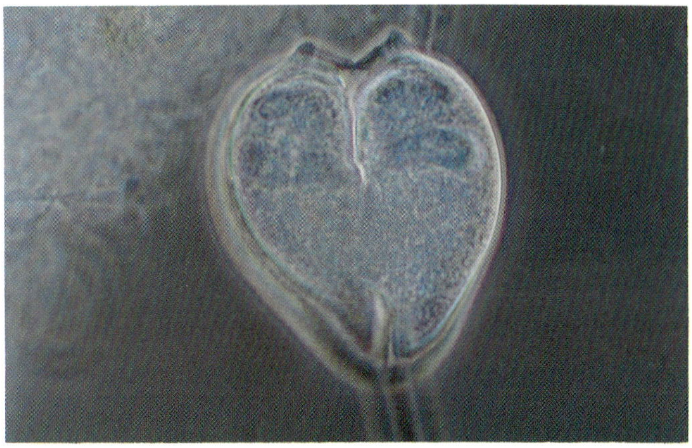

Figure 111. Dividing peritrich, from activated sludge, Cookeville, TN. Phase contrast. If it is a solitary species, one of the cells will swim away to settle elsewhere (**Figure 112**), while the other will retain the stalk. If it is a colonial species, the cells will split down to the stalk and remain connected to each other (**Figure 113**).

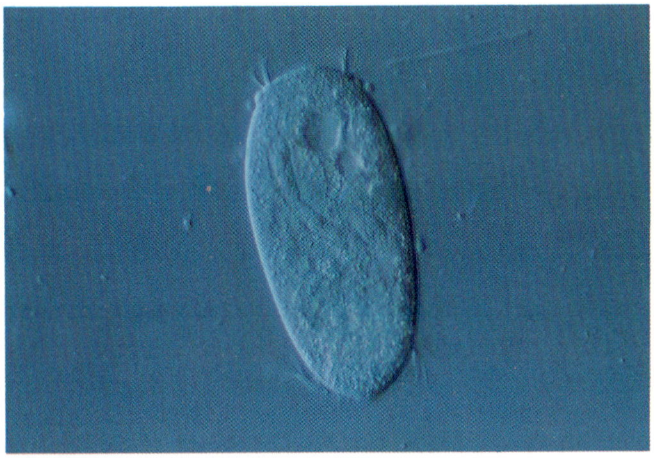

Figure 112. Motile telotroch larva produced from division of a sessile peritrich ciliate, 80 µm long, from clarifier walls, Cookeville, TN. Differential interference contrast.

Figure 113

Figures 113, 114. Colonies of *Epistylis*, 80 to 110 µm above stalk, from bottom of clarifier, Cookeville, TN. Differential interference contrast. Several colonial peritrichs occur commonly in wastewater treatment plants. *Opercularia* has oral cilia raised up on a small peg-like projection above the cell body. *Carchesium* has stalks with contractile filaments, but in which the filament stops just short of the junction between stalks, so each cell contracts independently. *Zoothamnium* contains contractile filaments connecting all members of a colony, so that all cells contract together. *Epistylis* lacks contractile filaments in the stalks.

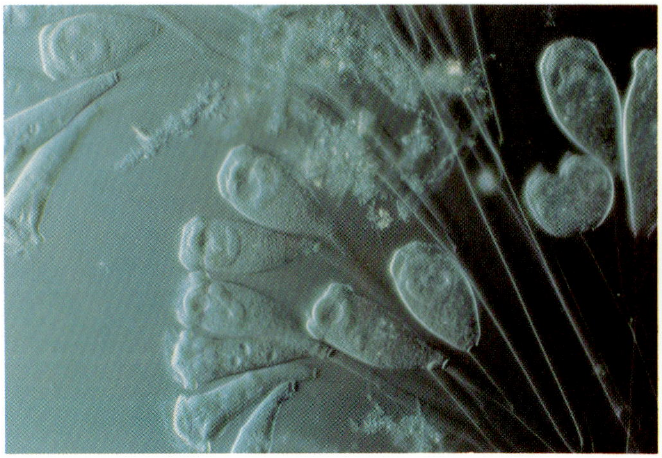

Figure 114

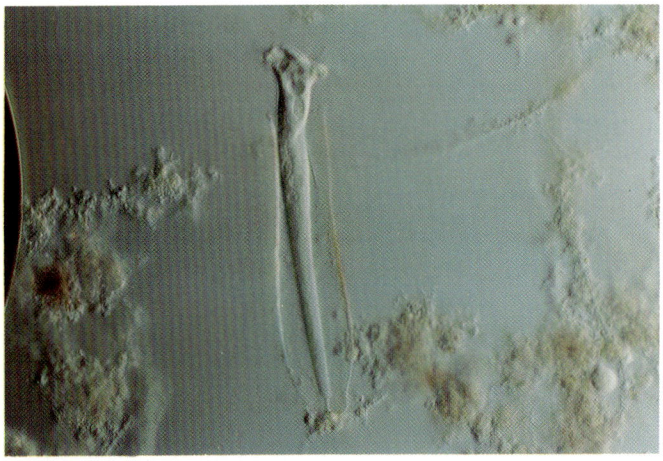

Figure 115. *Thuricola*, a peritrich ciliate, 220 µm long, from aeration basin, Murfreesboro, TN. Differential interference contrast. This peritrich secretes a lorica.

Figure 116. Food vacuoles forming in *Thuricola*. Cilia can be seen hanging down into the top vacuole in which bacteria are collected. Fine surface striations can be seen further down on the cell. Subtle surface markings of this sort are important in the taxonomy of *Vorticella*, and this makes identification to species very difficult.

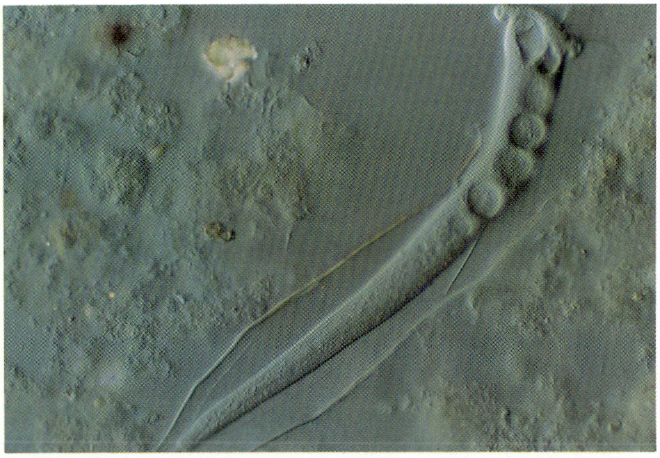

Figure 117. A slightly different optical section through the same specimen in **Figure 116**. Below the food vacuoles, a valve can be seen extending from the cell to the lorica. *Thuricola* is distinguished from the genus *Vaginicola*, very common in wastewater treatment plants, by the existence of this valve.

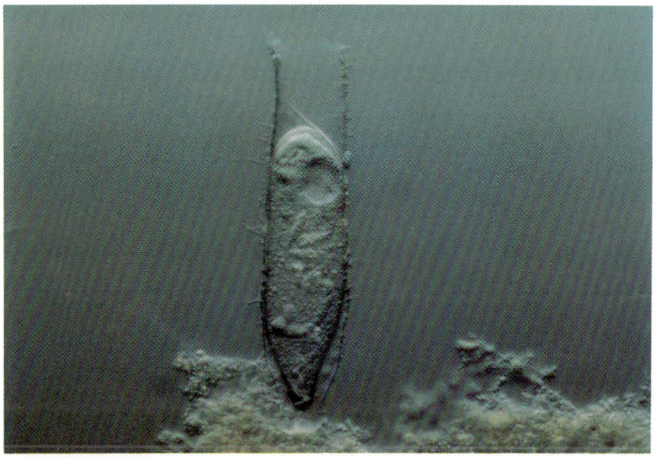

Figure 118. *Thuricola* contracted in its lorica, from activated sludge, Murfreesboro, TN. Differential interference contrast. The valve closes above the cell like a trap door.

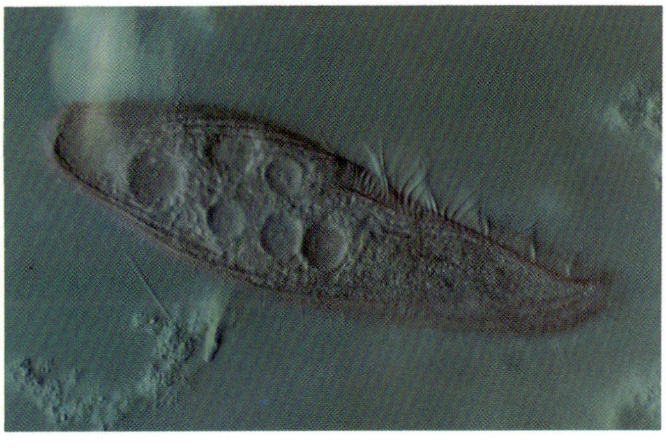

Figure 119. *Blepharisma*, 170 µm long, from final clarifier, Cookeville, TN. Differential interference contrast. This ciliate has well-developed oral cilia (on the top, right portion of the cell) used for filtering bacteria from the water. It also has uniform rows of somatic cilia used for swimming. Some species of *Blepharisma* are naturally pink in color.

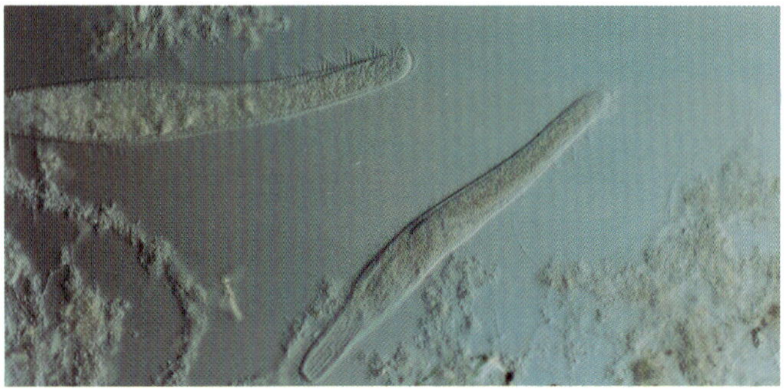

Figure 120

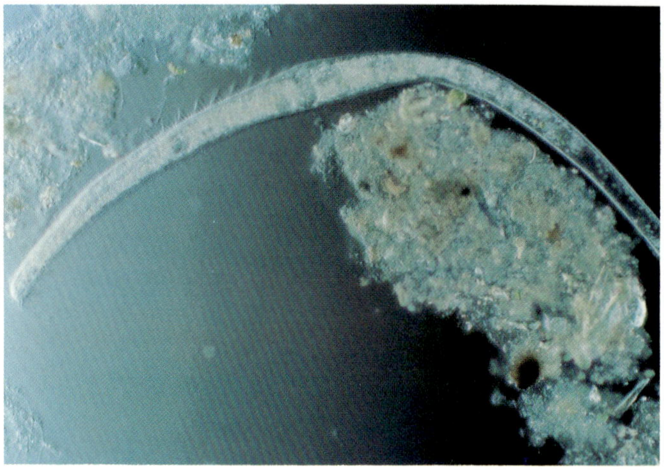

Figure 121

Figures 120, 121. *Spirostomum.* The ciliary pattern is the same as in *Blepharisma*, although the cells look quite different. Their worm-like bodies can reach over a half millimeter in length. The posterior end often has a clear vacuole. **Figure 120**, 250 μm long, from bottom of clarifier, Cookeville, TN. Differential interference contrast. **Figure 121**, 600 μm long, from bottom of clarifier, Cookeville, TN. Differential interference contrast.

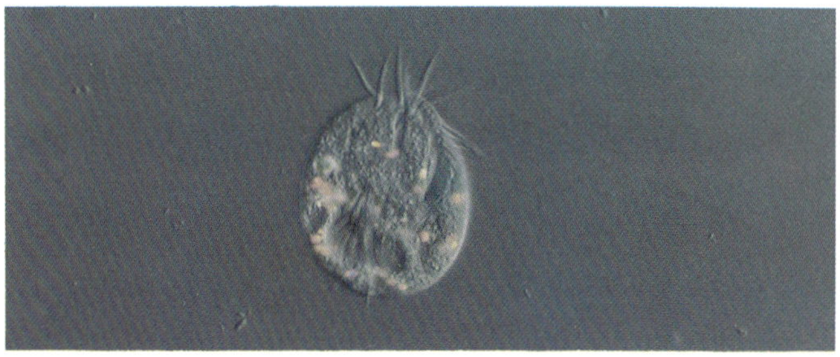

Figure 122

Figure 122. *Aspidisca*, a hypotrich ciliate, 52 μm long, from aeration basin, Murfreesboro, TN. Differential interference contrast. This is one of the most common ciliates in wastewater. The oral cilia are comparable in structure to those of heterotrichs like *Blepharisma* and *Spirostomum*, but somatic cilia are reduced and incorporated into compound spike-like organelles called cirri. Hypotrichs scuttle across surfaces using their cirri like legs. Several other hypotrich genera occur in these environments as well. They are distinguished by the number and arrangement of cirri.

Figures 123–132. Suctorians. These are ciliates that have no traces of cilia as an adult. They produce motile ciliated larvae, however. They have tentacles that attach to prey organisms, which are most commonly other ciliates. The prey cytoplasm is drawn down through the tentacles, and food vacuoles form at the base.

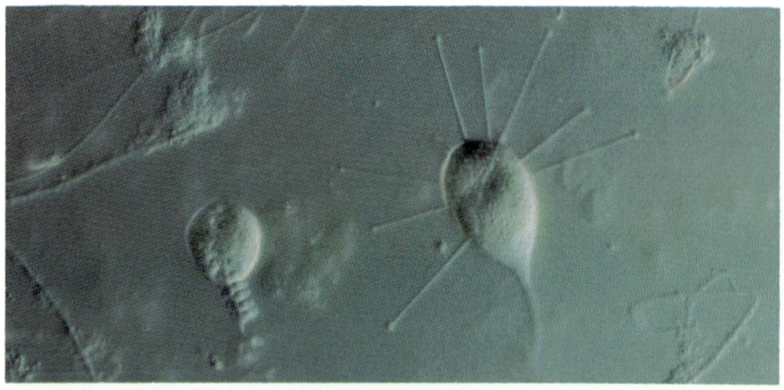

Figure 123

Figure 123. *Podophrya*, 50 μm long (excluding stalk and tentacles), from activated sludge, Murfreesboro, TN. Differential interference contrast. This is a stalked suctorian with tentacles distributed over the surface. A retracted *Vorticella* is next to the adult suctorian.

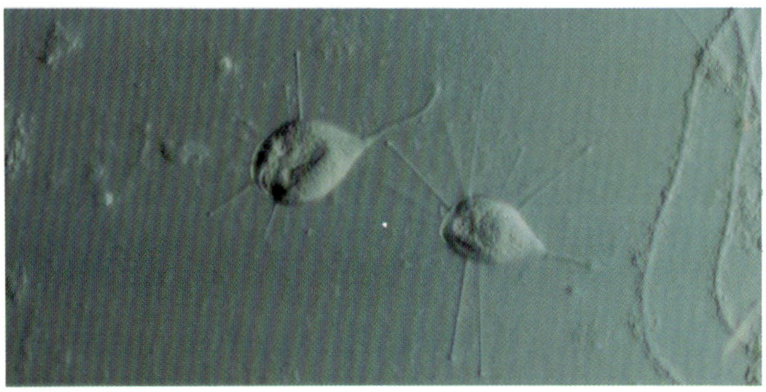

Figure 124. *Podophrya.* The ciliary field of the larva is beginning to form on the surface of the top individual.

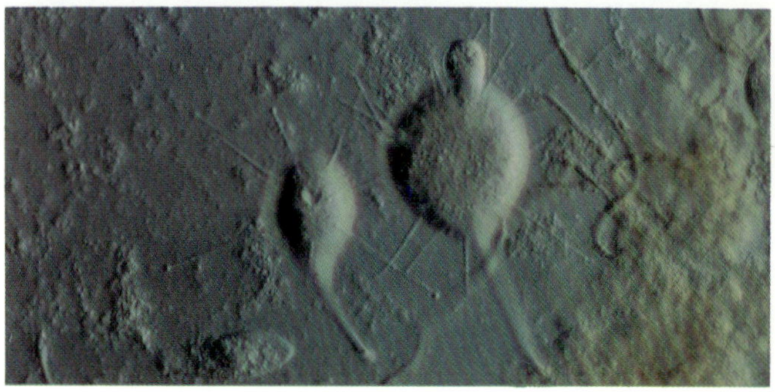

Figure 125. A ciliated larva of *Podophrya*, almost ready to detach and swim away.

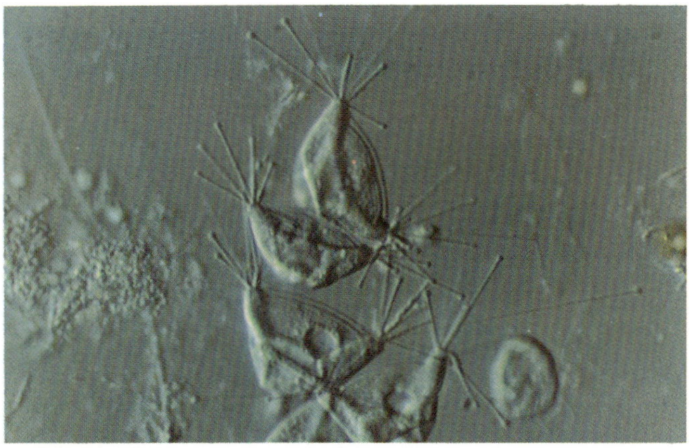

Figure 126. *Acineta*, 35 μm across, from activated sludge, Murfreesboro, TN. Differential interference contrast. The body is invested in a close-fitting lorica, and the tentacles are grouped into bundles.

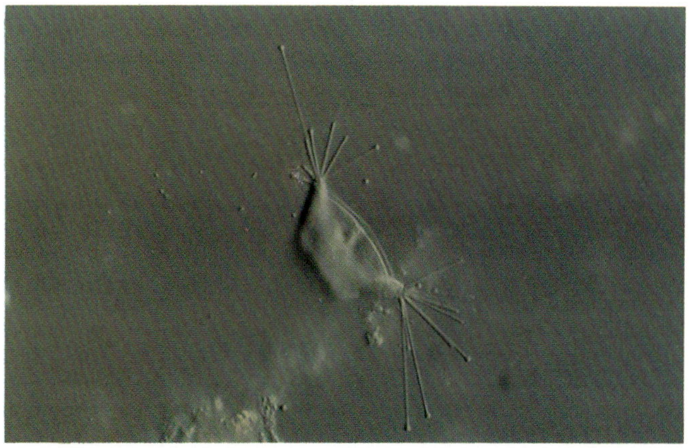

Figure 127. *Acineta*, 40 μm across, from activated sludge, Murfreesboro, TN. Differential interference contrast.

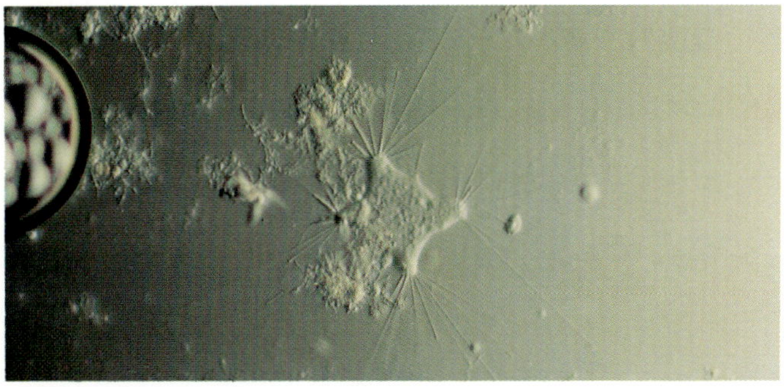

Figure 128. *Lernaeophrya*, 80 μm across, from activated sludge, Murfreesboro, TN. Differential interference contrast. This is a flattened, stalkless genus with bundles of tentacles distributed evenly around the body on finger-like projections.

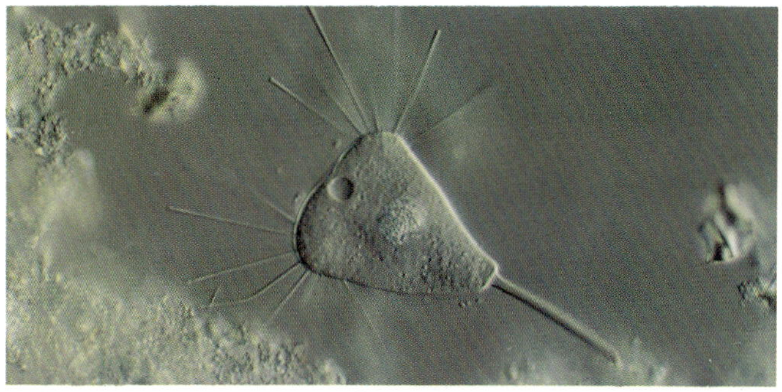

Figure 129. Unidentified suctorian, 55 μm long, excluding stalk, from activated sludge, Cookeville, TN. Differential interference contrast.

Figure 130. Unidentified suctorian, 110 μm, from aeration basin, Murfreesboro, TN. Differential interference contrast.

Figure 131. Unidentified suctorian, 60 μm long (lorica), from final clarifier, Murfreesboro, TN. Differential interference contrast.

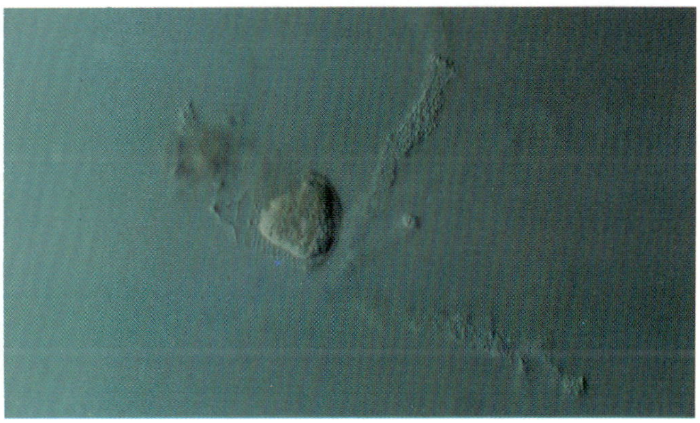

Figure 132. Unidentified suctorian, 70 μm long (lorica), from clarifier effluent, Cookeville, TN. Differential interference contrast. The branching structure next to the suctorian is the bacterium *Zoogloea*.

Invertebrates

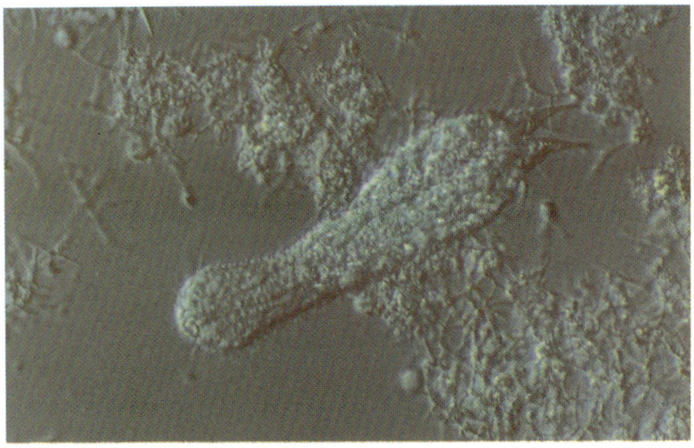

Figure 133

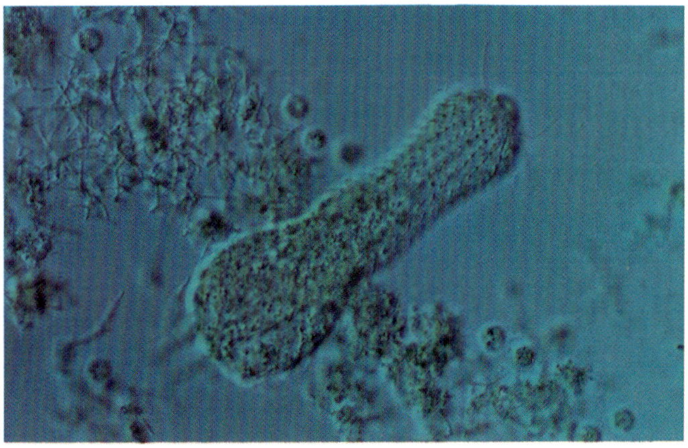

Figure 134

Figures 133, 134. *Chaetonotus* (gastrotrich), 120 μm long, from activated sludge, Murfreesboro, TN. Differential interference contrast. Gastrotrichs are metazoans that fall within the size range of large protozoans. The body is covered by a cuticle with spines (**Figure 134** shows the surface). Two projections are often at the posterior end. They also have cilia on the ventral surface of the head and trunk that help them glide along the surface. They graze on the substrate ingesting algae, small protozoa, bacteria, and detritus.

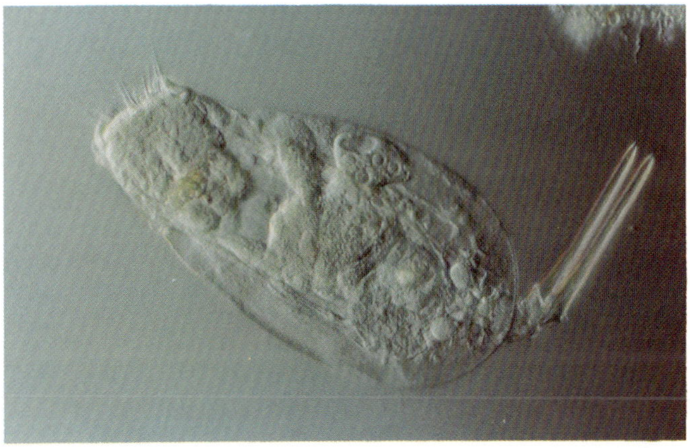

Figure 135. Ploimate rotifer, 200 μm long (including toes), from activated sludge, Murfreesboro, TN. Differential interference contrast.

Figure 136. The same specimen as **Figure 135** with foot and toes more contracted toward the body.

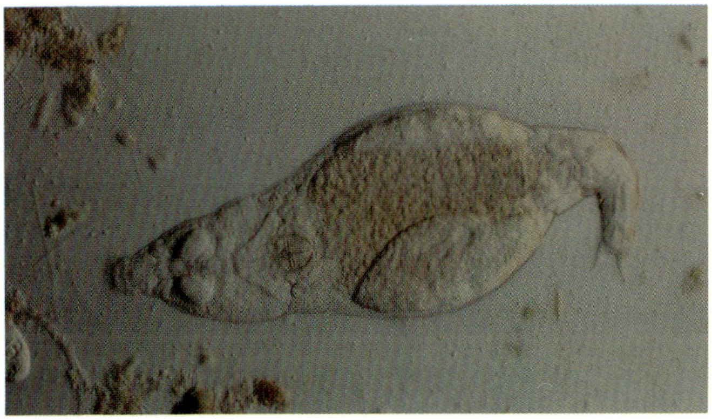

Figure 137. Bdelloid rotifer, unidentified species, 360 μm long, from trickling filter, Jefferson City, TN. Differential interference contrast. Two large ovaries and two toes are visible, as well as a ramate (grinding) mastax (jaws). These rotifers feed on suspended material brought to the jaws by ciliary motion of the cilia on the corona. The corona may be placed against a substrate, and ciliary motion may sweep material into the jaws.

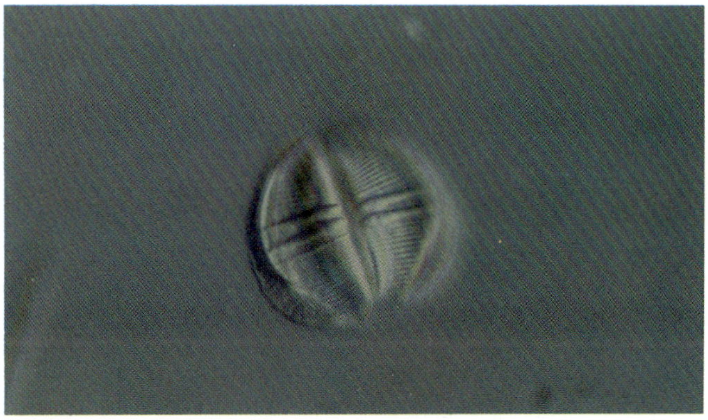

Figure 138. A ramate mastax of a bdelloid rotifer, 30 μm wide, from bottom of final clarifier, Cookeville, TN. Differential interference contrast.

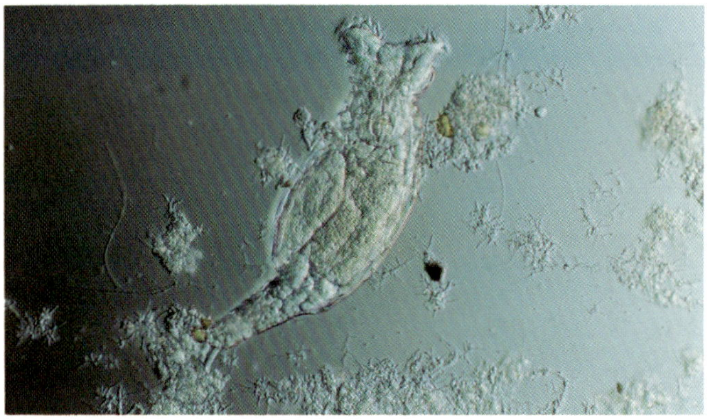

Figure 139

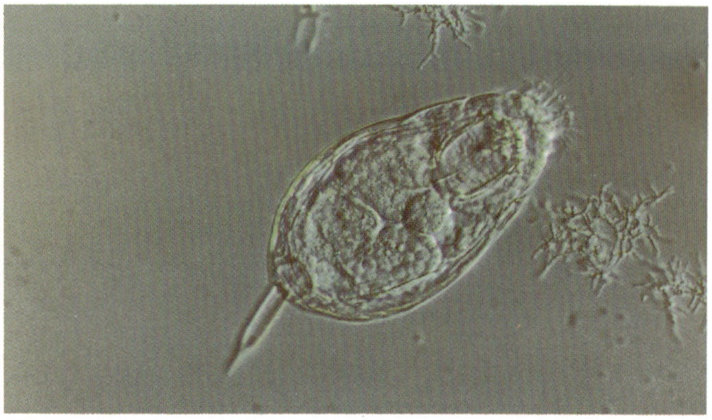

Figure 140

Figure 139. Bdelloid rotifer, 280 µm long, from aeration basin, Athens, TN. Differential interference contrast. This is another bdelloid rotifer with the same feeding habits as that in **Figure 137**. The corona with cilia is visible. It also has a ramate mastax.

Figure 140. *Lecane* sp. (=*Monostyla*), 130 µm long, from aeration basin, Athens, TN. Differential interference contrast. The most predominant family of rotifers in activated sludge is Lecanidae with *Lecane* as the most important genus. *Monostyla* is often united with *Lecane*. *Lecane* uses the foot primarily for locomotion. They have a malleate (chewing) mastax, but can also produce suction. The mastax can be applied directly to large pieces of material to be chewed before entering the stomach. Cilia are visible on the corona.

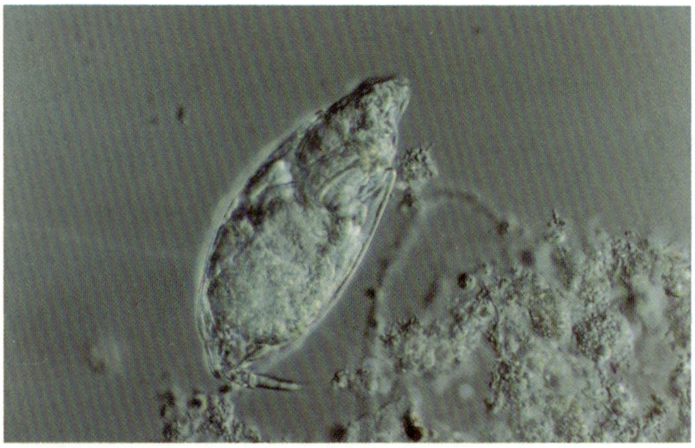

Figure 141. Another *Lecane* sp. (=*Monostyla*), similar to that in **Figure 140**, but more contracted, 135 µm long, from activated sludge, Murfreesboro, TN. Differential interference contrast. Cilia and the corona are not visible in the photograph.

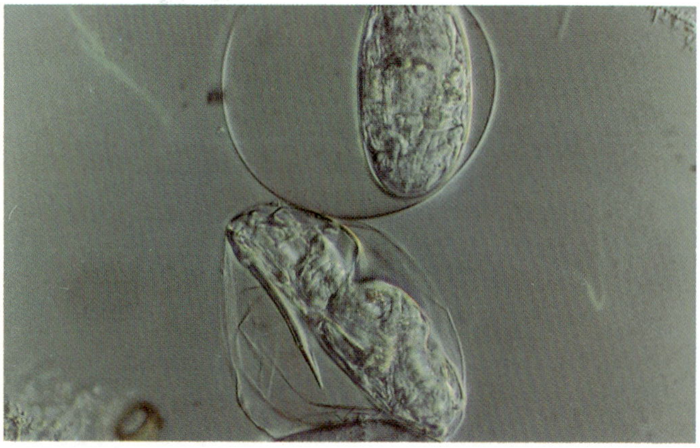

Figure 142. Rotifer eggs (probably *Lecane*), 100 µm long, in different stages of development, from clarifier effluent, Murfreesboro, TN. Differential interference contrast.

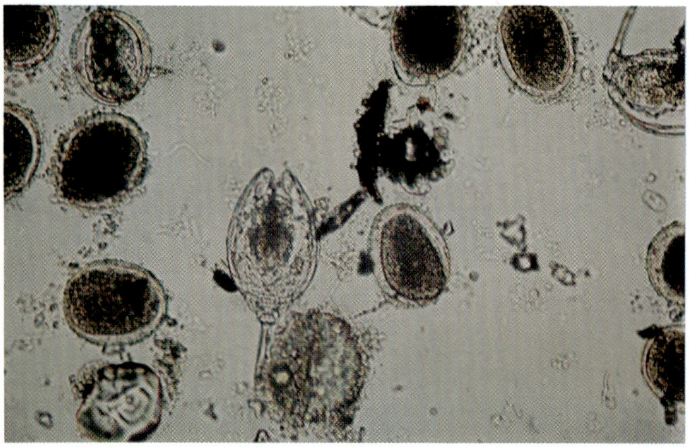

Figure 143. *Lecane* (= *Monostyla*) from activated sludge, New York. Brightfield.

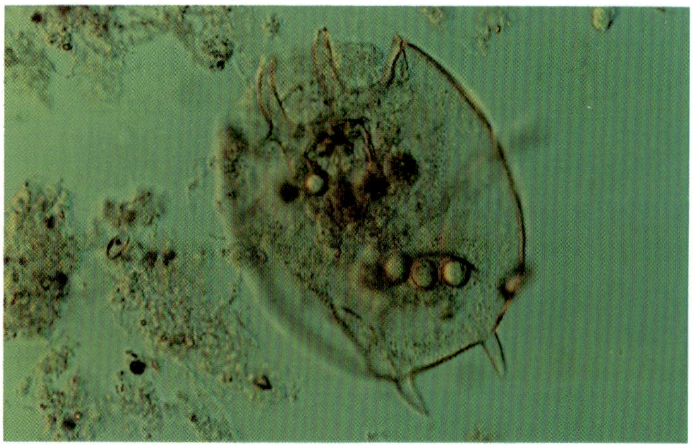

Figure 144. Lorica of *Platyias quadricornus* (rotifer), 200 µm long, from primary digester, Athens, TN. Differential interference contrast.

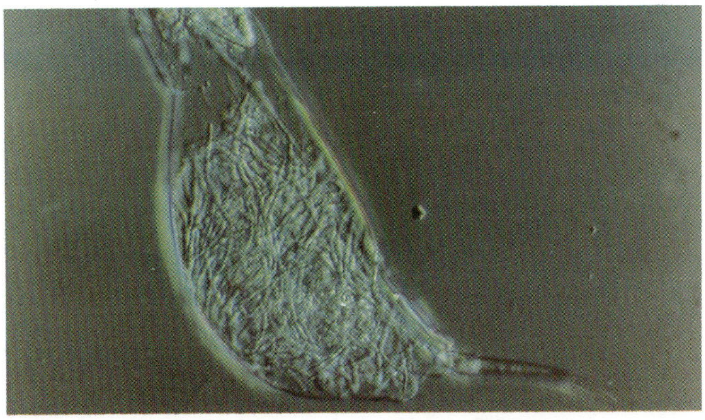

Figure 145. Unidentified dead rotifer, 270 µm long, filled with fungal filaments, from clarifier effluent, Murfreesboro, TN. Differential interference contrast.

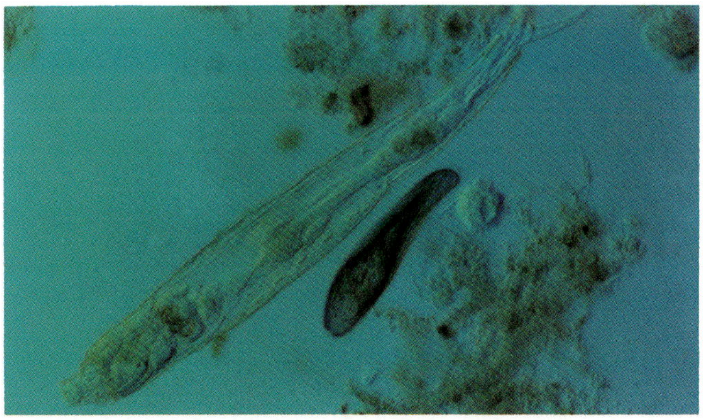

Figure 146. Unidentified rotifer, 480 µm long, and *Blepharisma* (ciliate), 150 µm long, from bottom of final clarifier, Cookeville, TN. Differential interference contrast.

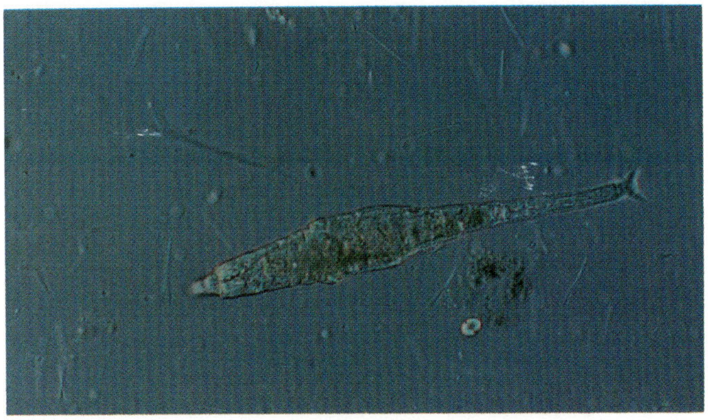

Figure 147

Figure 148

Figure 147. Unidentified rotifer, 410 µm long, from final clarifier, Cookeville, TN. Differential interference contrast.

Figure 148. Rhabditid nematode (roundworm), 1.3 mm long, from trickling filter material, Jefferson City, TN. Differential interference contrast. Nematodes are widely distributed in nature, but difficult to identify to species. The anterior end is truncate or bluntly rounded, and the posterior end tapers to a point. Most are colorless; any color is most likely from ingested food. Nematodes may be herbivores or carnivores. Some feed only on dead plant material, while others feed only on dead animal material. The food of some types of carnivorous species include protozoa, oligochaetes, rotifers, gastrotrichs, and other nematodes. Feeding is accomplished through a sucking action of the esophagus. They reach high population densities in wastewater treatment systems.

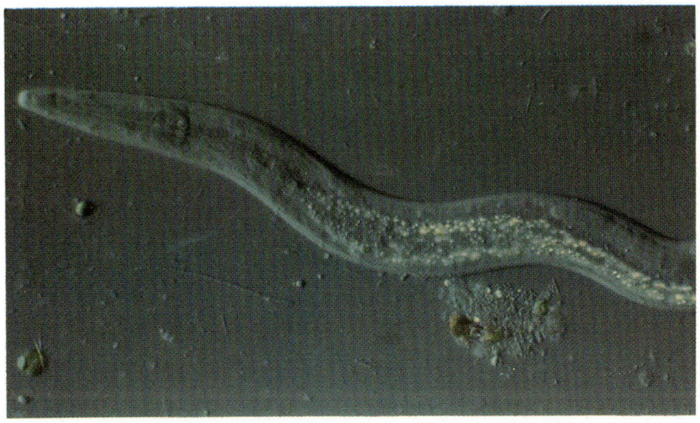

Figure 149. Another rhabditid nematode from trickling filter material, Jefferson City, TN. Differential interference contrast.

Figure 150. Nematode egg, 50 µm long, from trickling filter material, Jefferson City, TN. Differential interference contrast. Eggs are highly resistant to a lack of oxygen and can withstand several weeks of anaerobic conditions. They can also remain viable after repeated freezing and thawing.

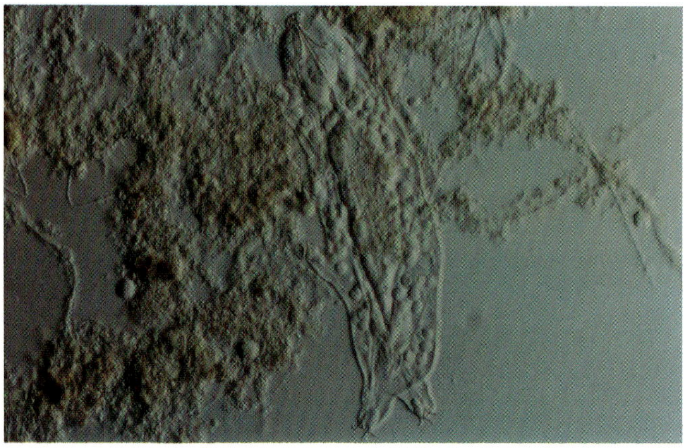

Figure 151

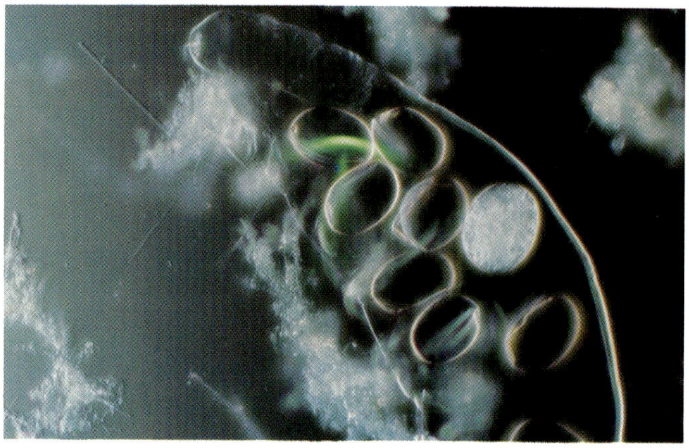

Figure 152

Figure 151. Tardigrade (water bear), probably *Hypsibius*, 250 μm long, from activated sludge, Murfreesboro, TN. Differential interference contrast. In nature, tardigrades are most commonly found in aquatic mosses and in the film of water in terrestrial mosses, liverworts, and angiosperms with forms that can retain water. They are encountered in some wastewater treatment plants. They slowly crawl among debris using the pairs of claws on each of eight legs to cling to the material. Feeding is accomplished by sucking produced by pharyngeal pumping action. They are mainly plant feeders, but some can feed on nematodes and rotifers.

Figure 152. Tardigrade exoskeleton containing eggs or remains of eggs from aeration basin, Murfreesboro, TN. Differential interference contrast.

Figure 153. Tartigrade eggs, 80 μm long, in exoskeleton, from aeration basin, Murfreesboro, TN. Differential interference contrast.

Figure 154. Statoblast of a bryozoan (rare in wastewater) from Baxter, TN. Differential interference contrast. Bryozoans have a plant-like appearance and form sessile colonies, the basic unit of which is a cylindrical zooid with tentacles.

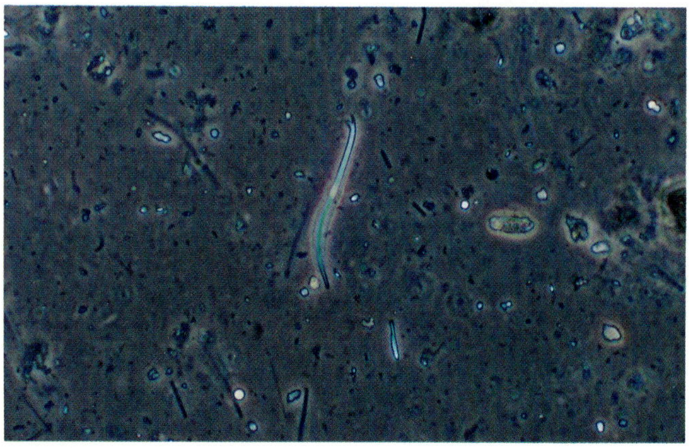

Figure 155. Seta (bristle-like projection) of an annelid worm, 40 μm long, from primary digester, Monterey, TN. Phase contrast. The seta is the highly refractile object with a forked end and bulge in the center. All other dark rods are large bacteria.

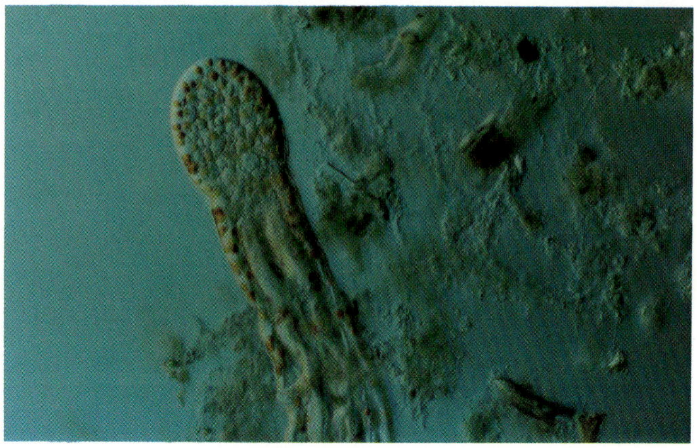

Figure 156. Anterior end of *Aeolosoma* sp. from bottom of final clarifier, Cookeville, TN. Differential interference contrast. This is an oligochaete annelid that feeds on particulate debris and microorganisms swept toward the mouth by cilia (not visible in the photograph) and also by suction feeding. This species has orange epidermal glands seen in the photograph. They reproduce asexually by budding.

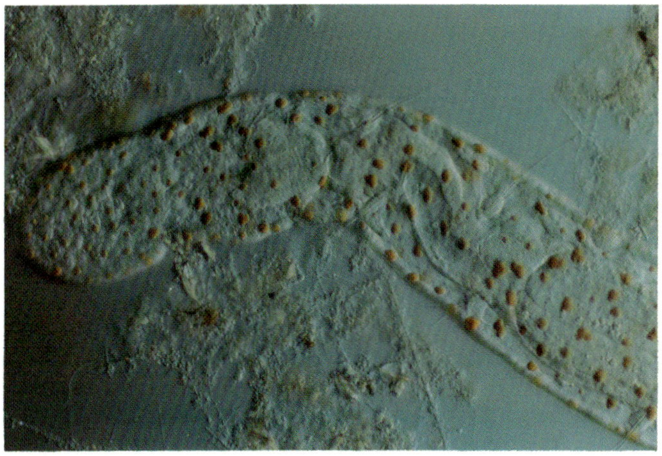

Figure 157. Another specimen of *Aeolosoma* from the bottom of the final clarifier, Cookeville, TN. Differential interference contrast. Bundles of setae (arrows) are visible in the photo.

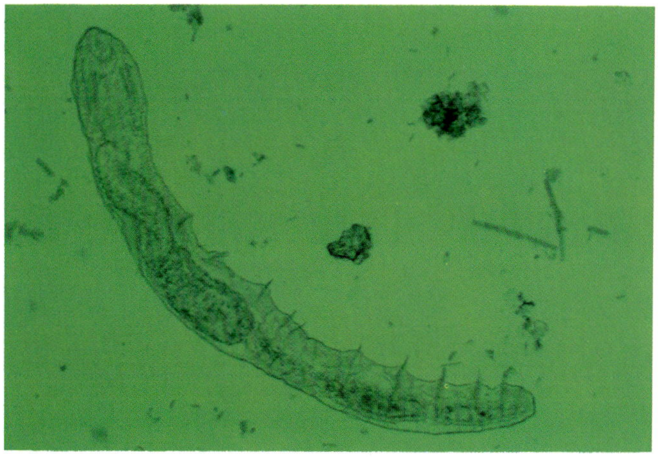

Figure 158. An oligochaete, possibly *Aeolosoma*, from bottom of final clarifier, Cookeville, TN. Brightfield with green filter. The intestinal tract can be seen running the length of the organism. Bundles of setae protruding from the side are also visible.

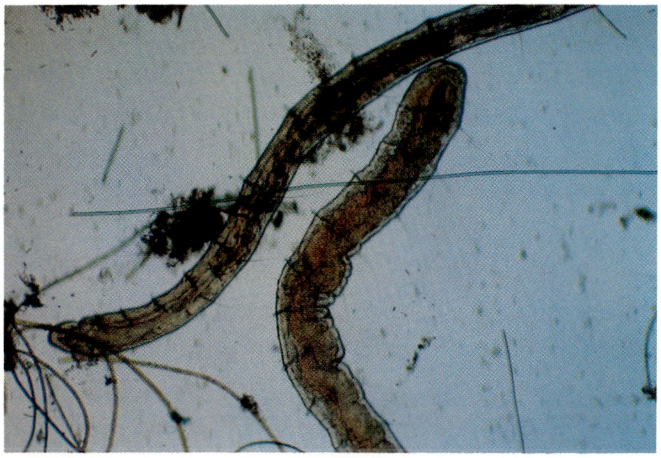

Figure 159. *Dero furcatus*, anterior, total length ≈8 mm, from final clarifier, Cookeville, TN. Brightfield. This is an oligochaete annelid that feeds much like an earthworm. They are often associated with filamentous algae (some can be seen in the photograph) and ingest algae, diatoms, and plant and animal detritus. They have ciliated anal gills on the posterior end (see **Figure 160**).

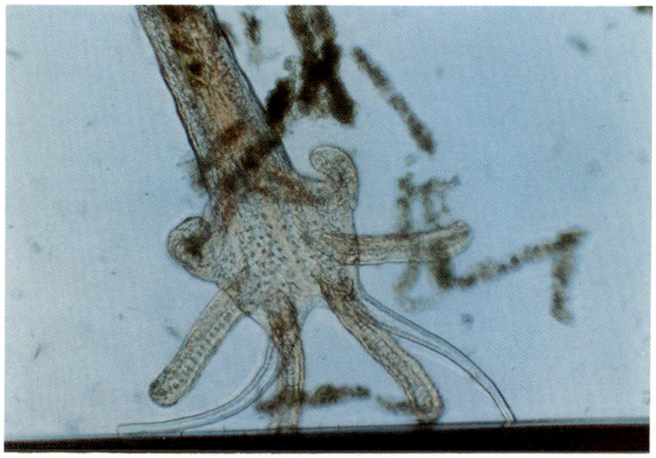

Figure 160. Posterior end of the same specimen of **Figure 159**. Material in the foreground may be fecal material from the annelid.

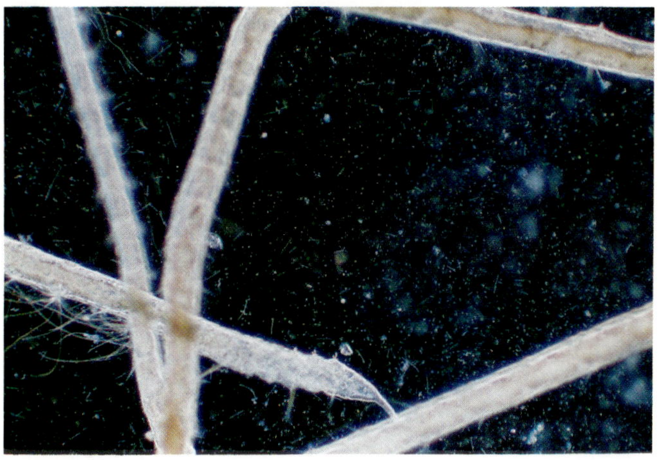

Figure 161. Anterior end of an unidentified oligochaete annelid from clarifier walls, Cookeville, TN. Darkfield. Note the proboscis-like projection of the prostomium (a half-segment at extreme anterior end of the body).

Figure 162. The gill-bearing posterior end of *Dero* sp. and a different genus of oligochaete from clarifier walls, Cookeville, TN. Darkfield.

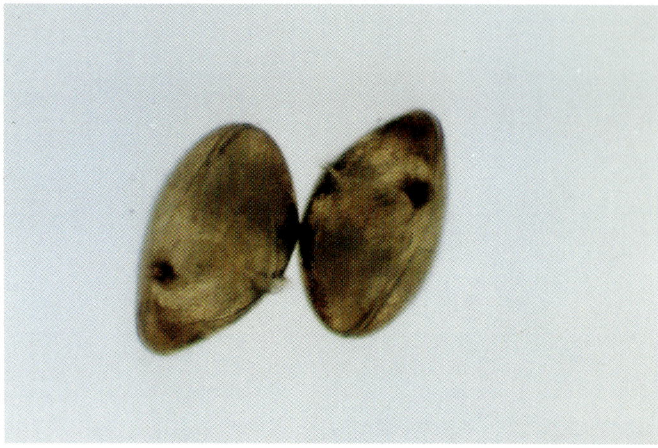

Figure 163. Ostracods, 320 µm long, among filamentous algae, from walls of final clarifier, Baxter, TN. Brightfield. Ostracods are arthropods that have a bivalve shell, and resembles clams or mussels. They feed primarily on bacteria, fungi, algae, and fine detritus.

Figure 164. Ostracod, 320 µm long. Setae of an appendage are visible sticking out of the valves. The body and appendages are not visible.

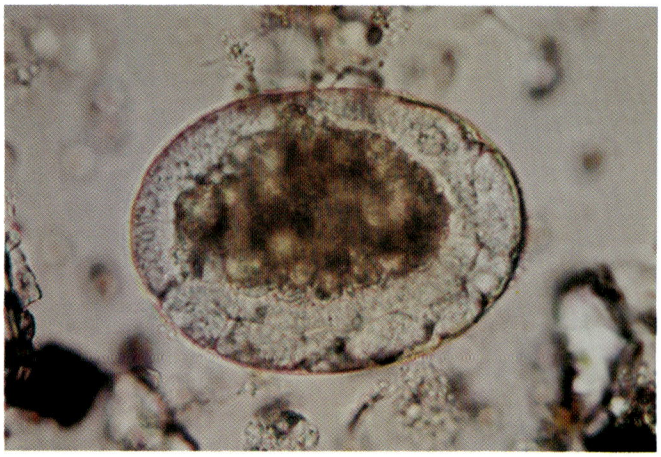

Figure 165. Mite egg from sludge, source unknown. Brightfield. Photograph taken by Dr. M. D. Little.

Parasites

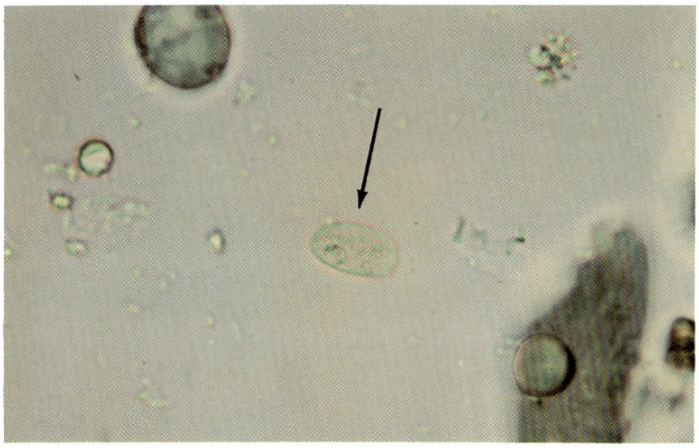

Figure 166. Cyst of *Giardia* sp. (arrow) ≈10 μm in length, from primary sludge, Alabama. These are parasitic flagellates that live in the small intestine of vertebrates. Resistant cysts, which serve to infect the next host, are passed out with feces.

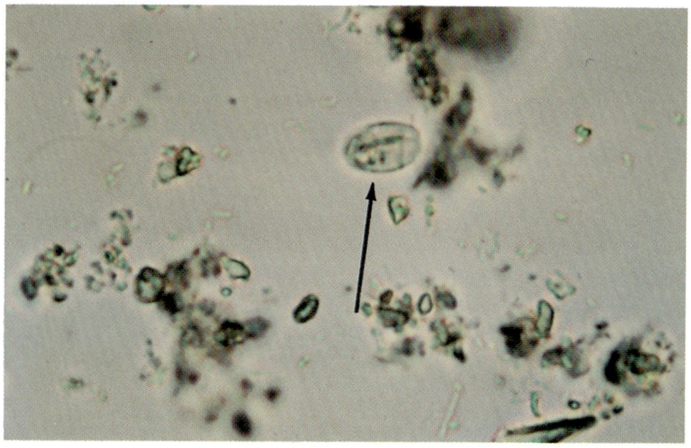

Figure 167. Same as **Figure 166** except dead.

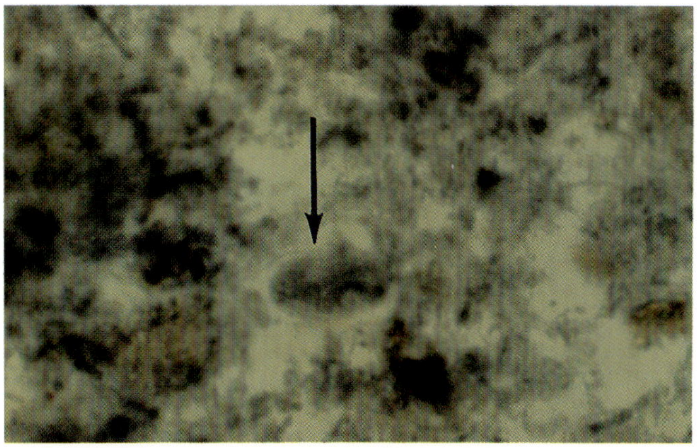

Figure 168. Cyst (arrow) of *Giardia intestinalis* (=*lamblia*), ≈12 μm in length, stained with hematoxylin. This species infects humans. The active flagellates in the intestine have two nuclei and eight flagella. After encystment, the nuclei divide, so that four nuclei are present in cysts. Several are in the plane of focus here. The other darkly staining structures in the cyst are microtubular bands and fragments of the sucking disc with which it attaches to the intestinal wall.

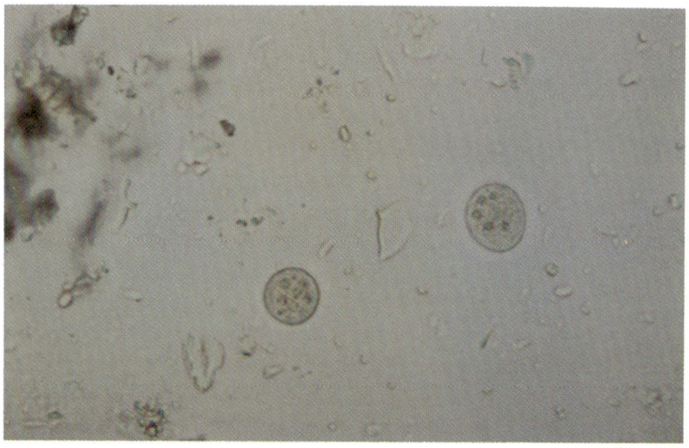

Figure 169. *Entamoeba coli* cysts, ≈10 μm in diameter, from sludge, Washington. These amoebae live as commensals in the intestine and eat bacteria without doing harm to their host. There are several rounds of nuclear division after encystment, so that mature cysts contain eight nuclei. They can be distinguished from *E. histolytica*, the causative agent of amoebic dysentery in humans, by the number of nuclei in cysts. *E. histolytica* cysts contain four nuclei.

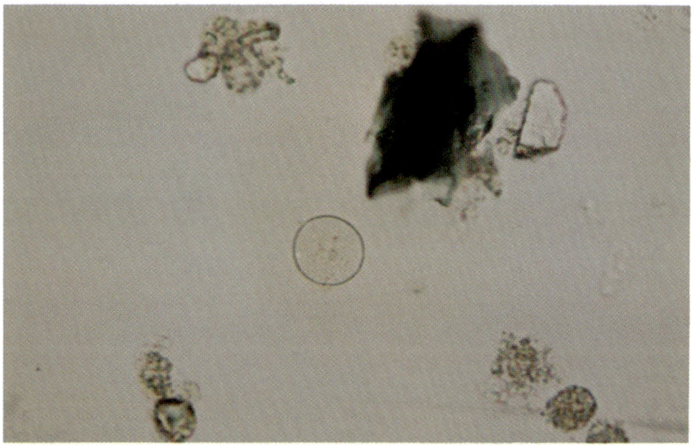

Figure 170. *E. coli*-like cyst, ≈12 μm in diameter, from primary sludge, New York.

NOTE: All photographs of parasites except for Figures 168 and 172 were contributed by Dr. M. Dale Little of Tulane University.

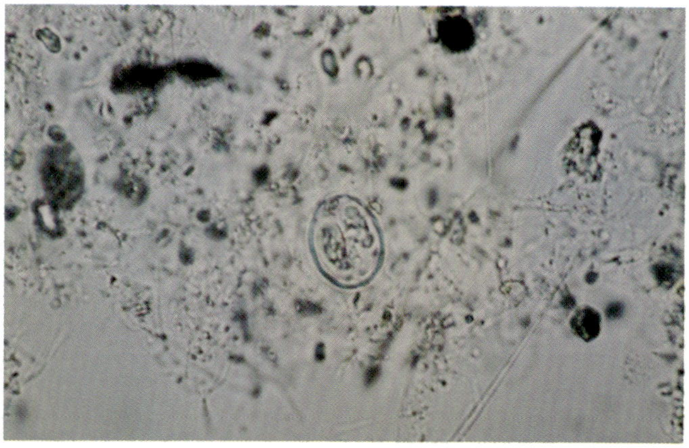

Figure 171. Coccidian oocyst, ≈12 μm long, from sludge drying bed, Alabama. When an oocyst is swallowed by an appropriate host, sporozoites inside will emerge in the intestine and begin multiplying in the cells of the intestinal epithelium. Many species of coccidia exist. Identification of this species is not possible from this photograph.

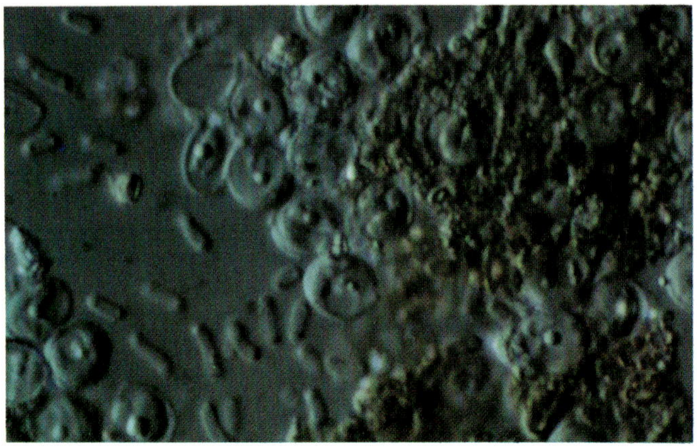

Figure 172

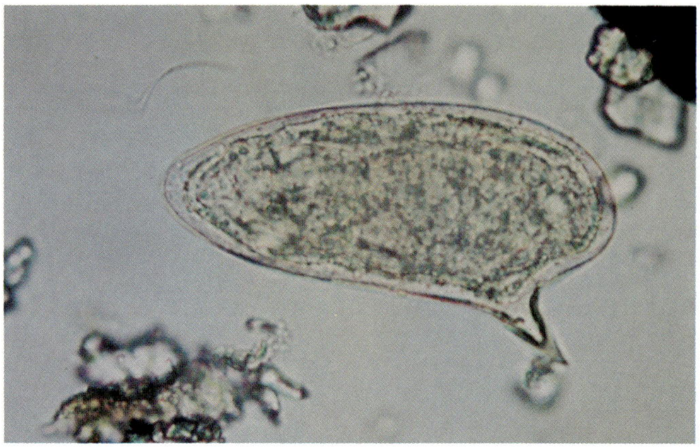

Figure 173

Figure 172. *Cryptosporidium parvum* oocysts (round structures) and sporozoites (smaller cylindrical structures with rounded ends) from cattle feces. Oocysts ≈5 μm in diameter; sporozoites ≈3 to 5 μm long. Differential interference contrast. This species is a coccidian parasite that can infect humans causing a self-limiting diarrhea. It can be more serious for immunocompromised individuals. Sporozoites would not survive outside the oocyst in wastewater. The oocysts of this photograph were placed in an excystation medium, so that sporozoites would be released.

Figure 173. Egg of the trematode *Schistosoma mansoni*, 125 μm long, from primary sludge, New York. The adult worms parasitize humans, living in the blood vessels surrounding the large intestine. Their eggs migrate through tissue to reach the lumen of the intestine, after which they are passed out in the feces. The larvae hatch from the egg in water and must find an aquatic snail of the genus *Biomphalaria* for the life cycle to continue. It bores into the snail tissues and begins growing. It produces large numbers of cercariae (a motile swimming stage that leaves the snail) that penetrate the skin of humans wading in the water and then migrate to the blood vessels around the large intestine, where they reach maturity. It is common in Africa and the New World tropics. Although different sanitation standards and the absence of the appropriate snail host prevent transmission from occurring in the continental U.S., it is still found here, as it is brought in by travelers and immigrants.

Figures 174, 175, 176. Egg of the tapeworm *Diphyllobothrium latum*, 60 μm long. **Figure 174** from primary sludge, Minnesota; **Figure 175** from human feces, preserved; **Figure 176** from primary sludge, New York. The egg has an operculum on one end and a barely developed knob at the other, features that are most apparent in **Figure 175.** The swimming larva that emerges from the egg survives only if eaten by a copepod. It bores through the intestinal wall and takes up residence in the hemocoel. If the copepod is then eaten by a fish, the tapeworm larva migrates to the muscle and grows further. They can be seen as white blobs in the meat of freshly caught fish. The life cycle is completed when a predator consumes the fish; the tapeworm reaches maturity in the intestine of the predator. *D. latum* is a common human parasite throughout the world. In the U.S., it is most common in the Great Lakes area.

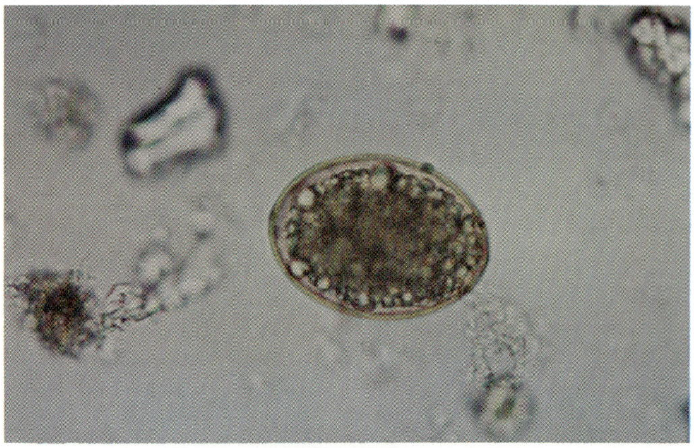

Figure 174

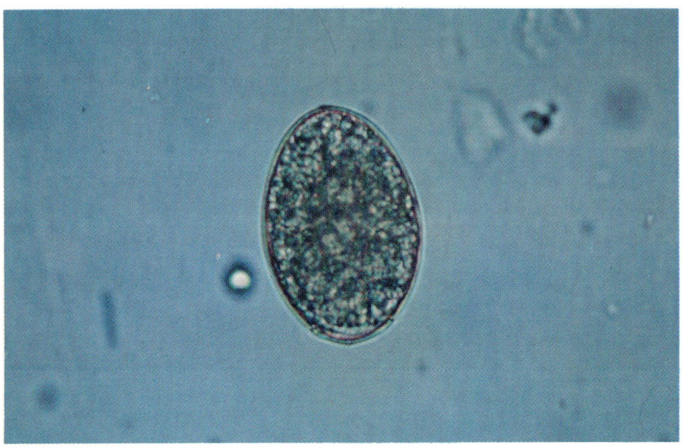

Figure 175

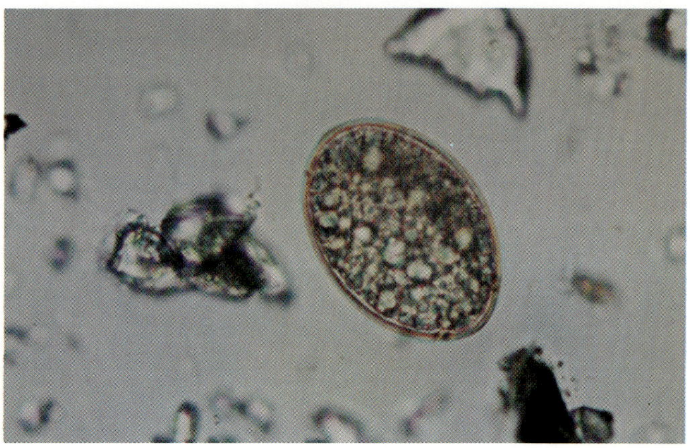

Figure 176

Figure 177. Egg of the tapeworm *Hymenolepis diminuta*, ≈90 μm long, from primary sludge, Washington. It will hatch after being ingested by a grain beetle, and the larvae will bore through the intestinal wall using the six hooks visible in the photograph. It will settle in the hemocoel and wait for a rat or human to ingest the beetle. After ingestion by a rat or human, the tapeworm will grow to maturity in the intestine.

Figures 178-184. Development of an *Ascaris* egg. The nematode *Ascaris lumbricoides* is a common human parasite that lives as an adult in the intestine. The eggs, about 60 μm long, are characteristically bumpy and brown. People become infected by ingesting eggs containing a young worm (**Figure 184**). Rather than remaining in the intestine, the worms migrate to the lungs. After crawling up to the throat from the lungs or being coughed up, the worms are reswallowed, and this time they stay in the intestine where they grow into adult worms. **Figure 178** from anaerobic digester, New York. **Figure 179** from sludge, source unknown. **Figure 180** from sludge drying bed, Alabama. **Figure 181** from primary sludge, New York. **Figure 182** from primary sludge, New York. **Figure 183** from activated sludge, Washington. **Figure 184** from sludge drying bed, Florida.

Figure 178

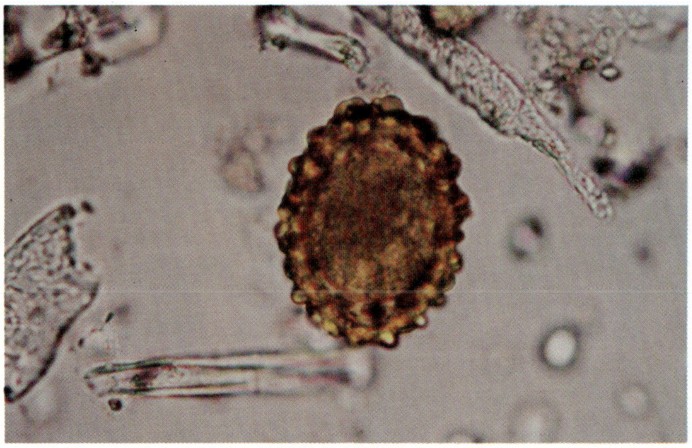

Figure 179

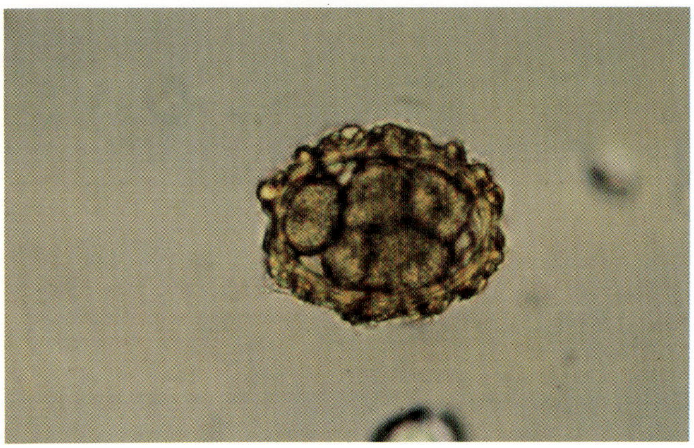

Figure 180

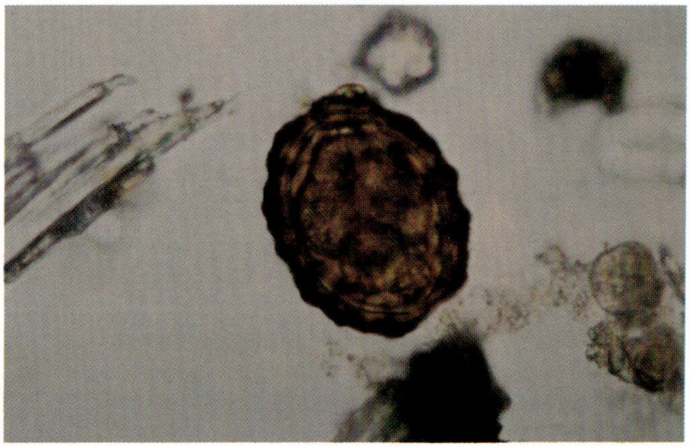

Figure 181

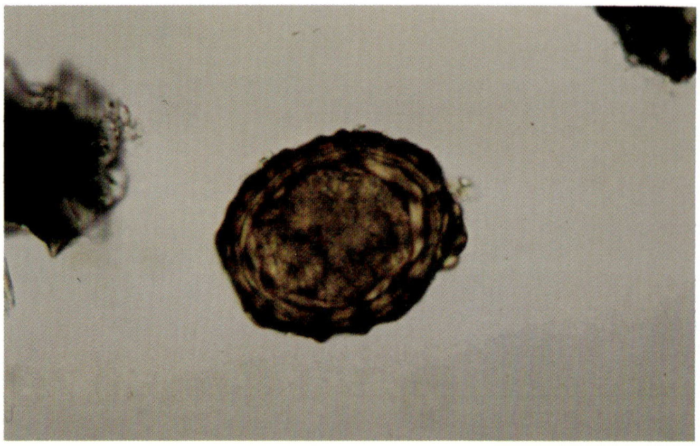

Figure 182

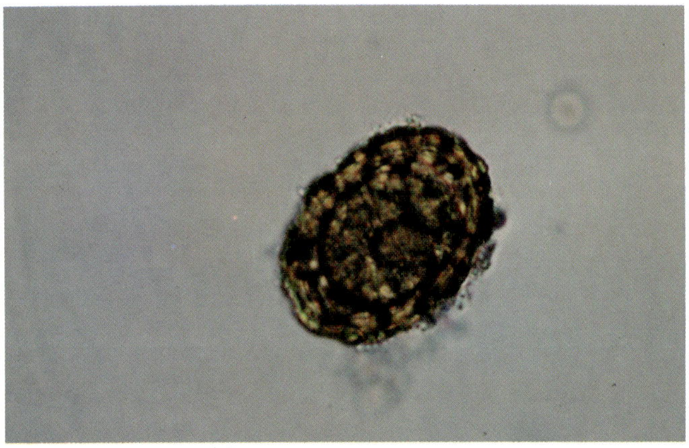

Figure 183

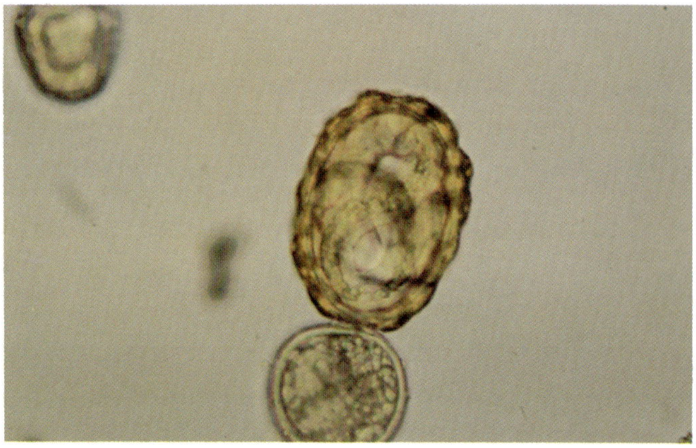

Figure 184

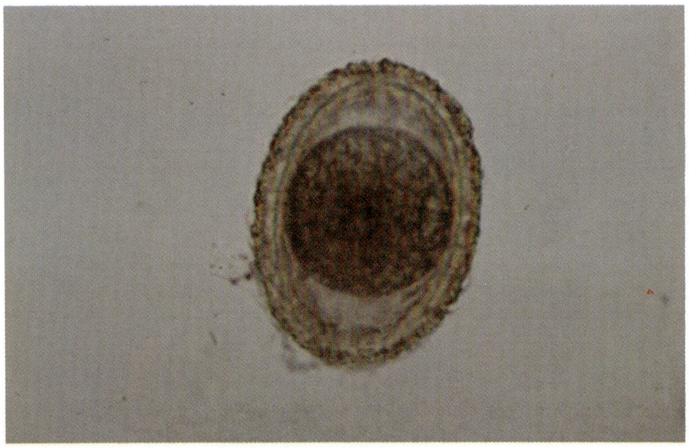

Figure 185. Egg of *Parascaris equorum* from anaerobic digester, Washington. This is an ascarid parasite of horses and has the same life cycle as *A. lumbricoides*.

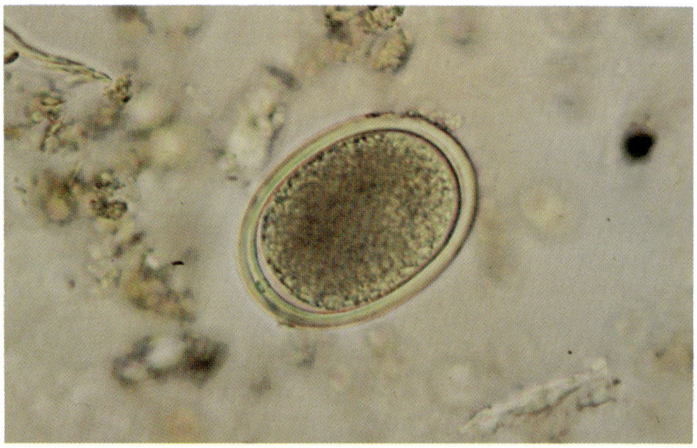

Figure 186. Nematode egg (*Ascaridia* sp. or *Heterakis* sp.) from poultry waste, Alabama.

Figures 187–192. Development of a *Toxocara canis* egg. This ascarid parasite of dogs has the same life cycle as that of *A. lumbricoides* when it infects puppies. In adult dogs, the larvae cannot complete migration to the lungs. They survive in the tissues for a while before dying. **Figure 187**, *T. canis* surface, from primary sludge, Minnesota. **Figure 188**, two-cell stage, from anaerobic digester, Washington. **Figure 189**, four-cell stage, from raw sludge, New York. **Figure 190**, eight or more cells, from sludge drying bed, Alabama. **Figure 191**, early larval stage, from primary sludge, New York. **Figure 192**, infective stage, source unknown.

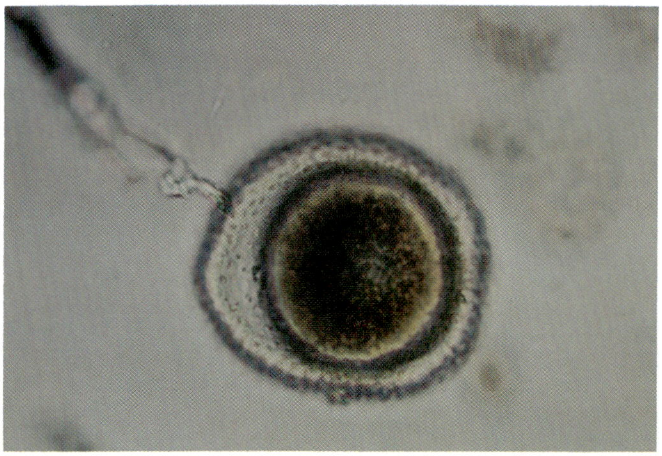

Figure 187

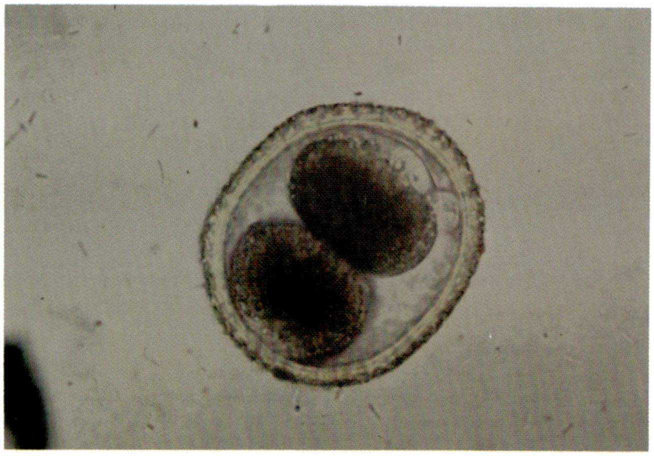

Figure 188

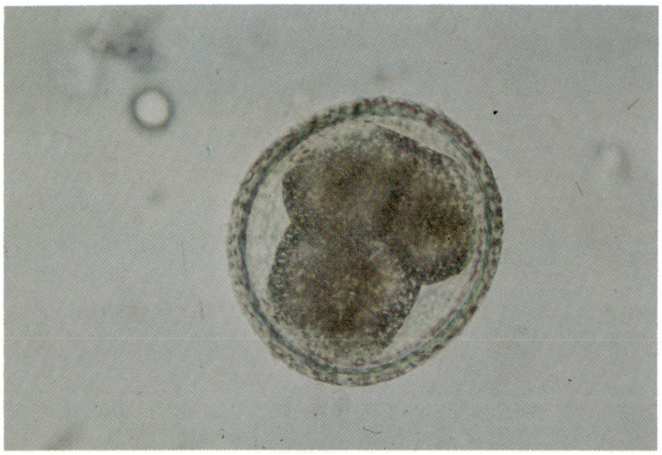

Figure 189

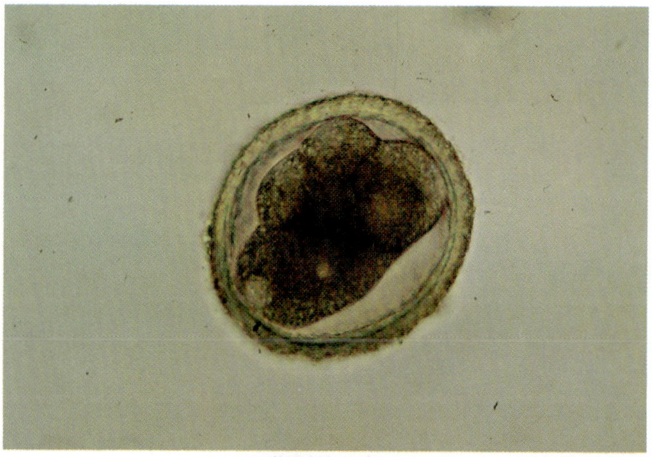

Figure 190

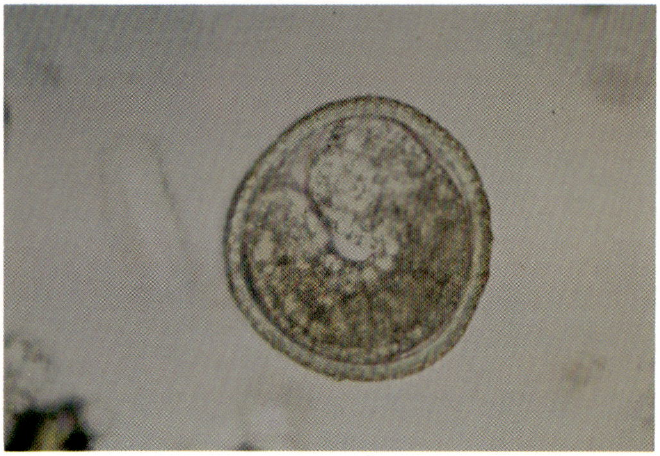

Figure 191

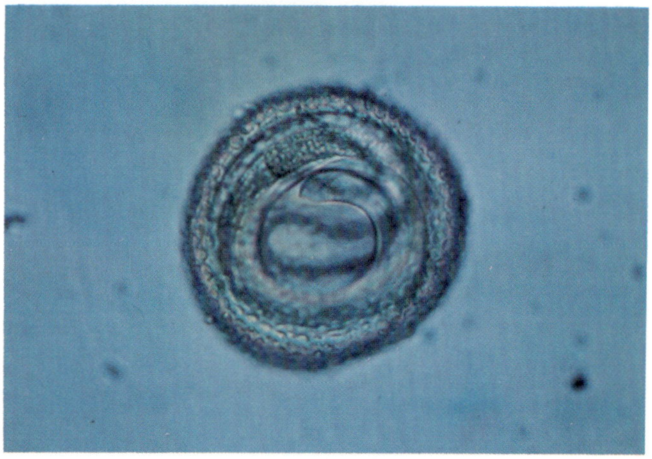

Figure 192

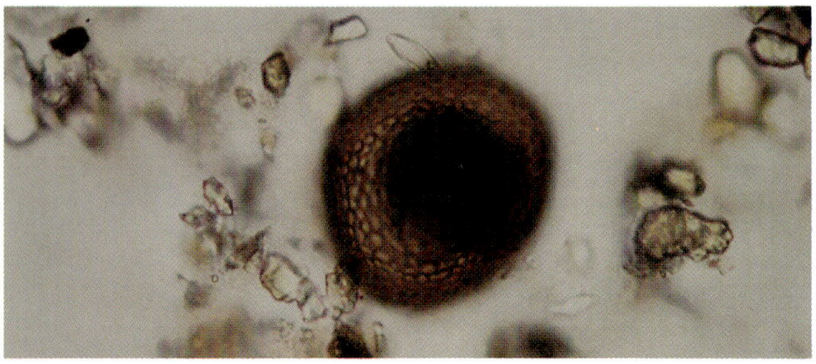

Figure 193

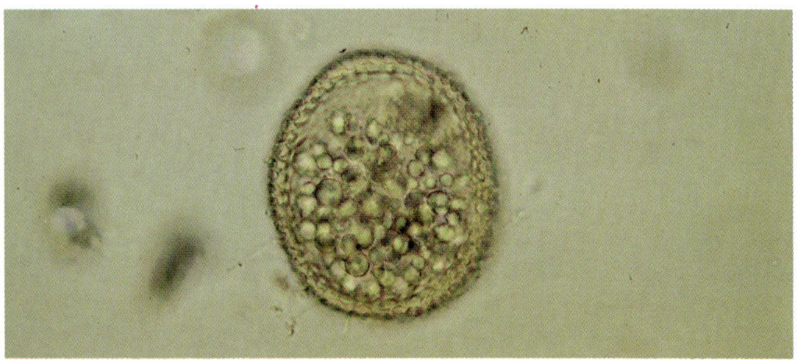

Figure 194

Figures 193, 194. Dead *T. canis* eggs. **Figure 193** from activated sludge, New York. **Figure 194** from drying bed, Alabama.

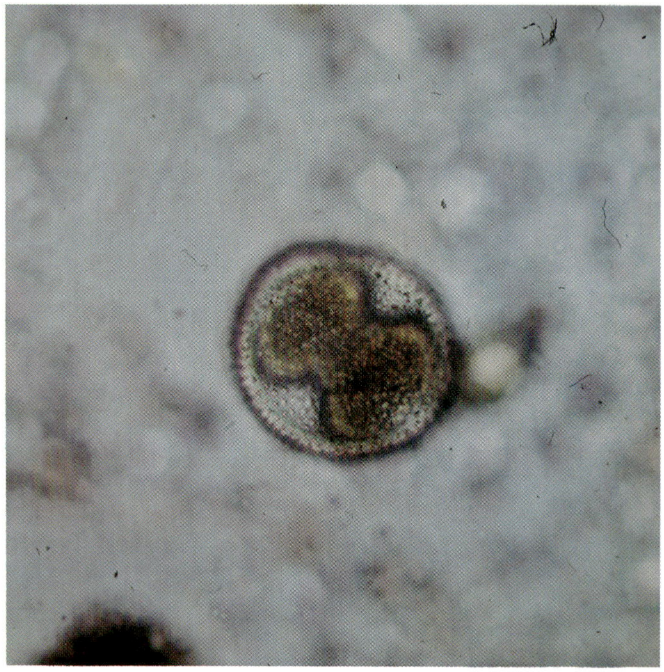

Figure 195. Egg of *Toxocara cati*, which infects cats, from Imhoff tanks, New York. Egg is in the two-cell stage. *T. cati* eggs are 80 to 90 μm in diameter.

Figures 196–199. Development of *Trichiuris trichiura* egg. This nematode parasite of humans is sometimes called a whip worm, because it has a long thin anterior end and a short fatter posterior end—the whip handle. Adults live in the intestine, and eggs are passed in feces. Humans become infected by ingesting eggs. **Figure 196**, two-cell stage, from aerobic sludge, Alabama. **Figure 197**, eight-cell stage, from anaerobic digester, New York. **Figure 198**, *T. trichiura* viable egg from primary sludge, New York. **Figure 199**, infective egg, source unknown.

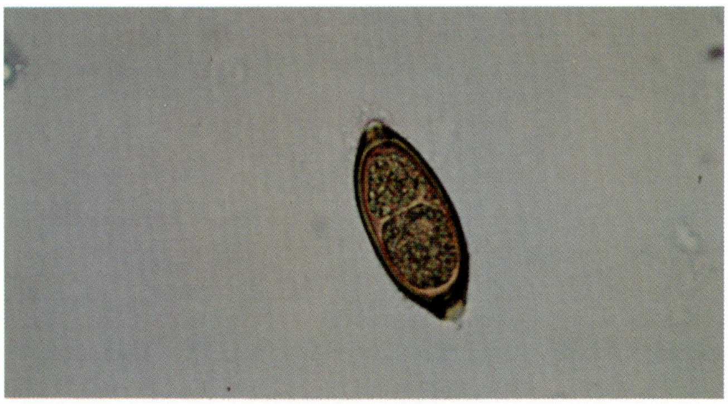

Figure 196

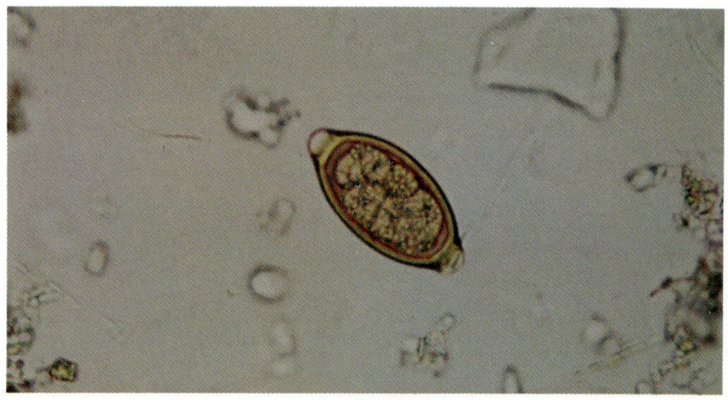

Figure 197

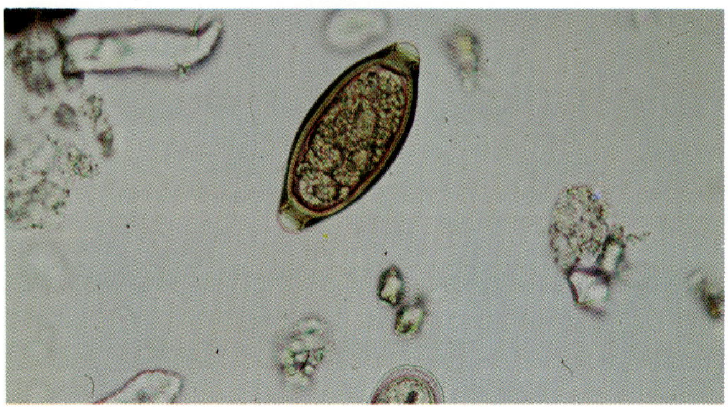

Figure 198

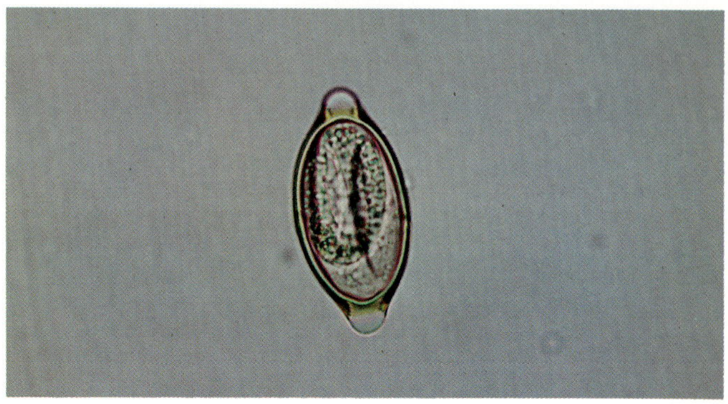

Figure 199

Figures 200–203. Eggs of *T. vulpis*, the dog whipworm. **Figure 200** from drying bed, Mississippi. **Figure 201**, two-cell stage, from anaerobic digester, New York. **Figure 202**, viable egg in the eight-cell stage, from primary sludge, New York. **Figure 203**, infectious stage, from drying bed, Florida.

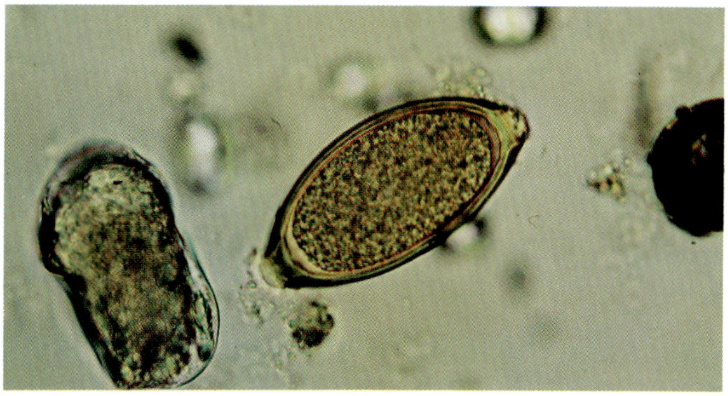

Figure 200

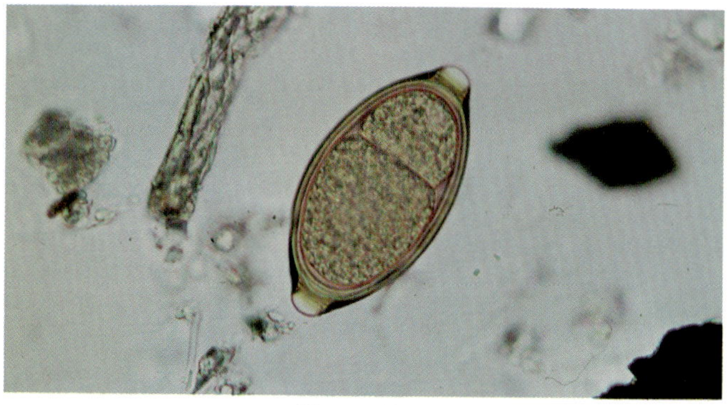

Figure 201

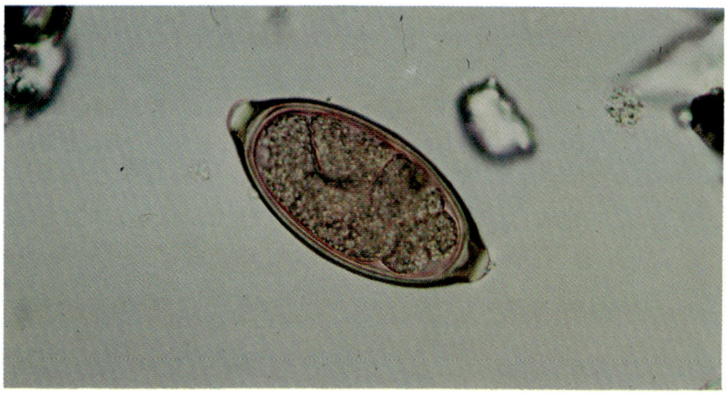

Figure 202

Figure 203

Figure 204. Egg of the rat nematode *Trichosomoides* from anaerobic digester, Washington.

Figure 205

Figure 206

Figures 205, 206. Unidentified nematode eggs from raw sludge, New York.

Pollen Grains

Pollen grains are occasionally encountered in wastewater samples and may confuse the inexperienced observer. Below are two types of pollen found in samples from Tennessee.

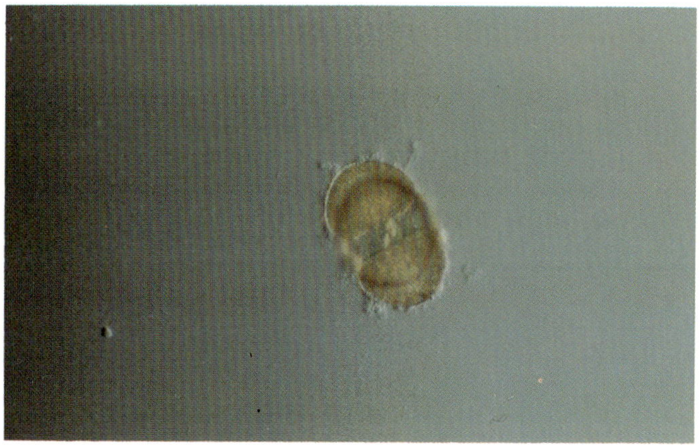

Figure 207. Pollen grain, 70 μm wide, from aeration basin, Murfreesboro, TN. Differential interference contrast.

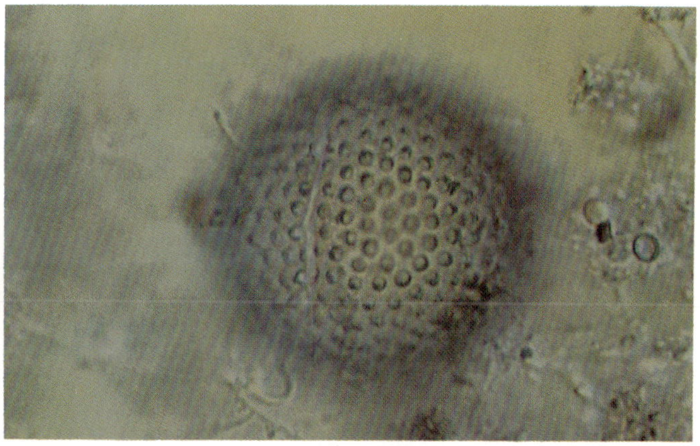

Figure 208

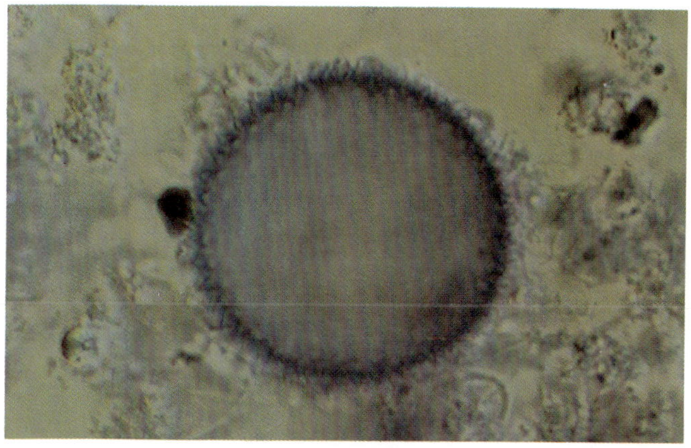

Figure 209

Figures 208, 209. Two different focal planes of a pollen grain, 70 μm in diameter, from return sludge, Baxter, TN. Differential interference contrast. **Figure 208** shows the golf-ball appearance of the surface.

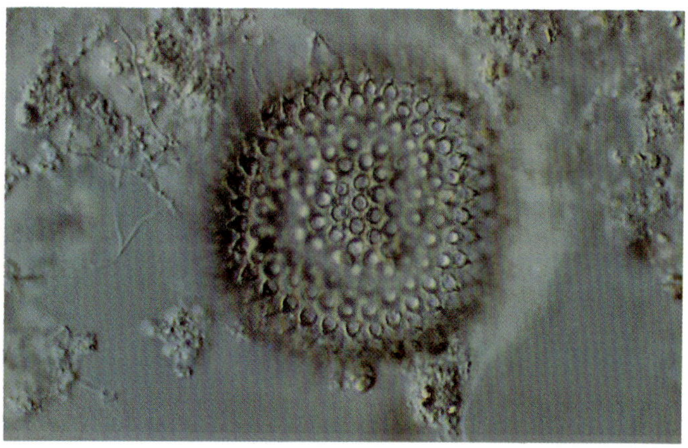

Figure 210. Another pollen grain from activated sludge, Baxter, TN. Differential interference contrast.

REFERENCES FOR BACTERIA

1. Bold, H. C. and M. J. Wynne. *Introduction to the Algae* (Englewood Cliffs, NJ: Prentice-Hall,1985).
2. Dugan, P. R. "The Genus *Zoogloea*," in *The Prokaryotes*, Vol. I., M. P. Starr, H. Stolp, H. G. Truper, A. Balows, and H. G. Schlegel, Eds. (Berlin: Springer-Verlag, 1981), pp. 764–770.
3. Goodfellow, M. and D. Minnikin. "The Genera *Nocardia* and *Rhodococcus*," in *The Prokaryotes*, Vol. II, M. P. Starr, H. Stolp, H. G. Truper, A. Balows, and H. G. Schlegel, Eds. (Berlin: Springer-Verlag, 1981), pp. 2016-2027.
4. Lechevalier, H. A. "Nocardioforms," in *Bergey's Manual of Systematic Bacteriology*, Vol. II, P. H. A. Sneath, N. S. Mair, and M. E. Sharpe, Eds. (Baltimore, MD: Williams & Wilkins, 1984), pp. 1458–1471.
5. Patterson, D. J. and S. Hedley. *Free-Living Freshwater Protozoa. A Color Guide* (Boca Raton, FL: CRC Press, 1992), p. 27.
6. Unz, R. F. "Genus IV. *Zoogloea* Itzigsohn 1868, 30," in *Bergey's Manual of Systematic Bacteriology*, Vol. I, N. R. Krieg, Ed. (Baltimore, MD: Williams & Wilkins, 1984), pp. 214–219.

REFERENCES FOR FLAGELLATES

1. Bold, H. C. and M. J. Wynne. *Introduction to the Algae* (Englewood Cliffs, NJ: Prentice-Hall, 1985).
2. Calaway, W. T. and J. B. Lackey. "Waste Treatment Protozoa. Flagellates," Florida Engineering Series No. 3, University of Florida, Gainesville, FL, 1962.
3. Curds, C. R. *An Illustrated Key to the British Freshwater Ciliated Protozoa Commonly Found in Activated Sludge* (London: Water Pollution Research Laboratory, Ministry of Tecnology, Water Pollution Research Technical Paper No. 12, Her Majesty's Stationery Office, 1969).
4. Curds, C. R. and H. A. Hawkes. *Ecological Aspects of Used-Water Treatment* (London: Academic Press, 1975).
5. Kudo, R. R. *Protozoology, 5th Edition* (Springfield, IL: Charles C Thomas, 1966).
6. Lee, J. J., S. H. Hutner, and E. C. Bovee. *An Illustrated Guide to the Protozoa* (Lawrence, KS: Society of Protozoologists, 1985).
7. Leedale, G. F. *Euglenoid Flagellates* (Englewood Cliffs, NJ: Prentice-Hall, 1967).
8. Martin, D. "Microfauna of Biological Filters," University of Newcastle Upon Tyne, Department of Civil Engineering, Bulletin 39, Oriel Press, Newcastle Upon Tyne, 1968.
9. Patterson, D. J. and S. Hedley. *Free-Living Freshwater Protozoa, A Color Guide* (Boca Raton, FL: CRC Press, 1992).

REFERENCE FOR NONFLAGELLETED ALGAE

1. Bold, H. C. and M. J. Wynne. *Introduction to the Algae* (Englewood Cliffs, NJ: Prentice-Hall, 1985).

REFERENCES FOR AMOEBAE

1. Curds, C. R. *An Illustrated Key to the British Freshwater Ciliated Protozoa Commonly Found in Activated Sludge* (London: Water Pollution Research Laboratory, Ministry of Tecnology, Water Pollution Research Technical Paper No. 12, Her Majesty's Stationery Office, 1969).
2. Curds, C. R. and H. A. Hawkes, *Ecological Aspects of Used-Water Treatment* (London: Academic Press, 1975).
3. Kudo, R. R. *Protozoology, 5th Edition* (Springfield, IL: Charles C Thomas, 1966).
4. Lee, J. J., S. H. Hutner, and E. C. Bovee. *An Illustrated Guide to the Protozoa* (Lawrence, KS: Society of Protozoologists, 1985).
5. Martin, D. "Microfauna of Biological Filters," University of Newcastle Upon Tyne, Department of Civil Engineering, Bulletin 39, Oriel Press, Newcastle Upon Tyne, 1968.
6. Ogden, C. G. and R. H. Hedley. *An Atlas of Freshwater Testate Amoebae* (Oxford: British Museum, Oxford University Press, 1980).
7. Page, F. C. *A New Key to Freshwater and Soil Amoebae* (Ambleside Cambria: Freshwater Biological Association, 1988).
8. Patterson, D. J. and S. Hedley. *Free-Living Freshwater Protozoa, A Color Guide* (Boca Raton, FL: CRC Press, 1992).

REFERENCES FOR CILIATES

1. Bick, H. *Ciliated Protozoa* (Geneva: World Health Organization, 1972).
2. Curds, C. R. *An Illustrated Key to the British Freshwater Ciliated Protozoa Commonly Found in Activated Sludge* (London: Water Pollution Research Laboratory, Ministry of Tecnology, Water Pollution Research Technical Paper No. 12, Her Majesty's Stationery Office, 1969).
3. Curds, C. R. and H. A. Hawkes. *Ecological Aspects of Used-Water Treatment* (London: Academic Press, 1975).
4. Kudo, R. R. *Protozoology, 5th Edition* (Springfield, IL: Charles C Thomas, 1966).
5. Lee, J. J., S. H. Hutner, and E. C. Bovee. *An Illustrated Guide to the Protozoa* (Lawrence, KS: Society of Protozoologists, 1985).
6. Martin, D. "Microfauna of Biological Filters," University of Newcastle Upon Tyne, Department of Civil Engineering, Bulletin 39, Oriel Press, Newcastle Upon Tyne, 1968.
7. Patterson, D. J. and S. Hedley. *Free-Living Freshwater Protozoa, A Color Guide* (Boca Raton, FL: CRC Press, 1992).

REFERENCES FOR INVERTEBRATES

1. Pennak, R. W. *Fresh-water Invertebrates of the United States* (New York: John Wiley & Sons, 1989).
2. Doohan, M. "Rotifera," in *Ecological Aspects of Used-Water Treatment*, C. R. Curds and H. A. Hawkes, Eds. (London: Academic Press, 1975).

REFERENCE FOR PARASITES

1. Schmidt, G. D. and L. S. Roberts. *Foundations of Parasitology* (St. Louis, MO: C. V. Mosby, 1977).

Glossary

Actinomycetes: Bacteria (usually soil bacteria) with some fungal-like characteristics, such as branching filaments that may resemble fungal hyphae.

Angiosperm: A flowering plant.

Apical: At the tip or peak.

Autoradiography: A technique in which an X-ray photograph is made by exposing film to radioactive materials.

Bdelloid rotifer: A type of rotifer with creeping, leech-like movement.

Brightfield: The use of direct light on a subject under the microscope resulting in a bright background.

Caudal: At or near the tail end.

Chemolithotrophic: Nutritional mode of bacteria that obtain energy by oxidizing inorganic nitrogen or sulfur compounds or methane or methanol.

Chlorophyllous: Containing chlorophyll.

Commensals: Organisms in a relationship with others, in which one species benefits from the relationship and the other is unaffected.

Corona: Disc-like ciliated anterior end of a rotifer.

Cyrtophorid: A type of ciliate with an oral apparatus consisting of a "basket" of microtubules.

Cytopharyngeal: Pertaining to the oral apparatus of protozoa, usually containing microtubules that draw food vacuoles into the cytoplasm.

Desmid: A unicellular green alga belonging to the order Zygnematales. They are grouped into two different families depending on their cell wall organization and mode of reproduction.

Epifluorescence microscopy: A technique that uses short wavelength light, an excitation filter, a beam splitter, and a barrier filter. The objective acts as the condenser. A specimen is illuminated with light of a particular wavelength, and the fluorescing light of a longer wavelength passes back through the objective and barrier filter to the eye piece.

Euglenoid: A type of flagellate, usually with a striated surface. Some have chloroplasts.

Fluorochromes: Dyes that fluoresce when illuminated with specific wavelengths of light.

Gamete: In plants and animals, a reproductive cell that is haploid.

Graticule: A grid in the eye piece of a microscope.

Helminth: A nematode, cestode, or trematode worm.

Hemocoel: The cavity carrying circulatory fluid in insects and crustaceans.

Heterotrophic: Refers to a mode of nutrition in which the nature of the carbon source is pre-formed organic material.

Holdfast: An attaching structure.

Imhoff tank: A two-story tank in which sedimentation is accomplished in the upper compartment and digestion occurs in the lower compartment.

Isogamous: Morphologically indistinguishable gametes.

Liverwort: Any of a class of bryophyte, which form moss-like mats in moist environments.

Lobose: Broad pseudopods.

Lorica: A housing around an organism. It may be made of organic or inorganic matter.

Metazoan: Multicellular animals.

Microthoacid: Ciliates with reduced perioral ciliature, in terms of numbers of cilia, and a reduced number of rows of somatic cilia. The body is usually flattened.

Microtubular: Consisting of structures made of the protein tubulin.

Monopodial: Having one pseudopod.

Mycelium: A network of hyphae as in fungi.

Oligochaete: A segmented worm lacking a definite head and having relatively few body bristles. They are found mainly in moist soil and freshwater.

Oocyst: A thick-walled cyst produced by some parasitic protozoa of the Class Sporozoa.

Operculum: A plug blocking the opening of some protozoan cysts; the lid or flap of some flatworm eggs.

Parthenogenetic: A mode of reproduction in which a gamete has the capacity to form a new organism without sexual union.

Pathogen: A disease-causing organism.

Pellicle: The complex outer layer of a protozoan cell, such as that covering ciliates and euglenoids.
Peptidoglycan: A compound that gives rigidity to prokaryotic cell walls, essential elements of which are alternating residues of N-acetylglucosamine and N-acetylmuramic acid, tetrapeptides (groups of four amino acids with alternating L and D configurations), and peptide bridges between tetrapeptides. Many variations within this framework are found among bacteria.
Phase contrast: Microscopy technique that uses special phase rings and objectives to enhance contrast between materials. Light waves become out of phase creating an image with more contrast than can be seen with brightfield microscopy.
Plastids: Chloroplasts, bodies containing pigments, and sometimes pyrenoids in which photosynthesis occurs.
Pleurostome: A ciliate with a slit-like mouth opening at what appears to be a lateral position on the body.
Ploimate rotifer: Common rotifers with a single ovary and nonramate mastax, frequently with a lorica.
Prokaryotes: An organism without membrane-bound nuclear material, having 70S type ribosomes, and one large circular DNA molecule.
Prostome: A ciliate with the mouth at the anterior end of the cell.
Pseudopods: A temporary cytoplasmic extension of the cell used for locomotion.
Pyrenoids: A center for starch formation within plastids.
Rhabditid: A nematode having a terminal pharyngeal bulb, pore-like sense organs, and no caudal glands.
Saprotrophic: Feeding on dissolved organic nutrients.
Scuticociliates: Small, common ciliates characterized by possession of a comma or whip-like pattern of basal bodies in the oral area during cell division.
Siliceous: Composed of silica.
Somatic: Of the body.
Sporozoites: Infectious form of coccidia formed within a cyst.
Suctorian: Ciliates that possess cilia only in the larval stage and feed as adults by using tentacles.
Telotroch: Ciliated dispersal stage of stalked peritrichs.
Tests: Shell-like "house" of organisms such as some amoebae.
Trichocysts: Bodies within ciliates such as *Paramecium*, which may be ejected under adverse conditions or for feeding in some species. They have the appearance of tiny spears.
Truncate: With a flattened end.
Ventral: At or near the belly or underside.

Index of Genera

Genera are indexed by Figure number

(Numbers in parentheses indicate captions in which genera are mentioned but not photographed)

Acineta *126, 127*
Aeolosoma *156–158*
Amphileptus *86, 87*
Arcella *70–73*
Ascaris *178–184*
Aspidisca *122*
Blepharisma *119, 146*
Bodo *25*
Calyptotricha *107*
Carchesium *(113, 114)*
Centropyxis *74*
Chaetonotus *133, 134*
Chilodonella *91–97*
Chroococcus *14*
Cinetochilum *106*
Cladophora *41*
Closterium *54, 55*
Coleps *83*
Colpidium *98, 99*
Coscinodiscus *(58)*
Cryptosporidium *172*
Cyclidium *105*
Dero *159, 160, 162*
Diatoma *61*
Difflugia *(76, 77)*
Diphyllobothrium *174–176*
Entamoeba *169*
Entosiphon *20, 21*
Epistylis *113, 114*
Euglypha *78–81*
Giardia *166–168*
Gloeocapsa *13*
Gomphonema *44*
Heterakis *(186)*
Hexamita *26–29*
Histiona *32*
Hymenolepis *177*
Hypsibius *177*
Lecane *140–143*
Lernaeophrya *128*
Litonotus *85*
Lyngbya *16*
Mastigamoeba *36*
Mastigella *(36)*
Melosira *(58)*
Microspora *50*
Monas *34*
Monostyla *140, 141, 143*
Mougeotia *51, 52*
Naegleria *(63)*
Nassula *88, 89*
Navicula *59*
Nitzschia *60, 62*
Nocardia *9*
Oedogonium *42–47*
Oicomonas (Oikomonas) *35*
Opercularia *(113, 114)*
Oscillatoria *15*
Paramecium *100–103*
Parascaris *185*
Peranema *19*
Platyias *144*
Podophrya *123–125*
Poteriodendron *30, 31*
Prorodon *82*
Pseudodifflugia *76, 77*
Rhizoclonium *38–40*
Scenedesmus *56, 57*
Schistosoma *173*
Spirillum *5*
Spirogyra *48, 49*
Spirostomum *120, 121*
Stigeoclonium *53*
Tetramitus *(63)*
Thecamoeba *65, 66*
Thuricola *115–118*
Toxacara *187–195*
Trachelomonas *17, 18*
Trachelophyllum *84*
Trepomonas *(26–28)*
Trichiuris *196–203*
Trichosomoides *204*
Trigomonas *(26–28)*
Trithigmostoma *91–97*
Vaginicola *(117)*
Vahlkampfia *(63)*
Vorticella *110*
Zoogloea *11*
Zoothamnium *(113, 114)*

NOTES

NOTES

NOTES

NOTES

NOTES

NOTES